7th International Conference on Engineering Mathematics and Physics (ICEMP 2018)

MATEC Web of Conferences Volume 186 (2018)

Prague, Czech Republic
15 – 18 January 2018

Editors:

H. Akca
R. Luca Tudorache
L. Abdullah

ISBN: 978-1-5108-6860-1

Printed from e-media with permission by:

Curran Associates, Inc.
57 Morehouse Lane
Red Hook, NY 12571

Some format issues inherent in the e-media version may also appear in this print version.

Printed by Curran Associates, Inc. (2018)

For additional information, please contact EDP Sciences – Web of Conferences at the address below.

EDP Sciences – Web of Conferences
17, Avenue du Hoggar
Parc d'Activité de Courtabœuf
BP 112
F-91944 Les Ulis Cedex A
France

Phone: +33 (0) 1 69 18 75 75
Fax: +33 (0) 1 69 28 84 91

contact-edps@webofconferences.org

Additional copies of this publication are available from:

Curran Associates, Inc.
57 Morehouse Lane
Red Hook, NY 12571 USA
Phone: 845-758-0400
Fax: 845-758-2633
Email: curran@proceedings.com
Web: www.proceedings.com

TABLE OF CONTENTS

PREFACE

It is my great pleasure to introduce you to the proceedings of 2018 The 7th International Conference on Engineering Mathematics and Physics (ICEMP 2018) which was held in Prague, Czech Republic on June 15-18, 2018.

ICEMP is an annual research conference aims to provide a platform for the scientists, scholars, engineers and students from the Universities all around the world and the industry to present ongoing research activities. The past 6 ICEMP conferences were held in Bangalore (India), Colombo (Sri Lanka), Hong Kong, Kuala Lumpur (Malaysia), and Dubai (UAE), and Cape Town (South Africa) respectively.

This volume consists of 20 selected papers from papers submitted to the conference from 16 countries and districts around the world including China, Russia, Mexico, India, Iran, Egypt, Thailand, Germany, Korea, Oman, Indonesia, Ukraine, Colombia, Turkey, Saudi Arabia, and France. The proceedings cover a wide range of topics including Applied Mathematics and Physics, Material Physics and Performance Analysis. Each contributed paper was strictly double blind peer-reviewed by international reviewers who were drawn from the organizing committee members and other experts. All the papers are evaluated through six aspects, i.e., relevance to the conference, originality, appropriateness of the research method, innovation, significance, English presentation. All the papers were also presented and examined during the conference orally or with posters.

We would like to thank all authors who have contributed to this volume and also to our chairs, keynote speakers, invited speakers, reviewers and other conference participants for their support to ICEMP 2018.

We hope ICEMP 2018 will generate many new innovative trends in the field and also all the authors and other interested readers will find the proceedings useful for their academic and research work. Last but not least, we are looking forward to meeting you at ICEMP 2019.

ICEMP 2018 Conference Chair
Prof. Haydar Akca
Abu Dhabi University, UAE

July 3, 2018

Statement of Peer review

In submitting conference proceedings to MATEC Web of Conferences, I certify to the Publisher that I adhere to the **Policy on Publishing Integrity** of the journal in order to safeguard good scientific practice in publishing

1. All articles have been subjected to peer review administered by the proceedings editors.
2. Reviews have been conducted by expert referees, who have been requested to provide unbiased and constructive comments aimed, whenever possible, at improving the work.
3. Proceedings editors have taken all reasonable steps to ensure the quality of the materials they publish and their decision to accept or reject a paper for publication has been based only on the merits of the work and the relevance to the journal.

Title, date and place of the conference

2018 the 7th International Conference on Engineering Mathematics and Physics
June 15-18, 2018
Prague, Czech Republic

Proceedings editor(s):

Prof. Haydar Akca
Prof. Rodica Luca Tudorache
Prof. Lazim Abdullah

Date and editor's signature 23 July 2018

 https://doi.org/10.1051/matecconf/201818601001

Study of Circulation Drive in High-Speed Rotating Flow Field

Xu Fan[1], and *Bo* Ran[1]

[1] *Gas Dynamics Sciences, Research Department, 300180 Tianjin, China*

Abstract. In the rotor with high speed, there is a certain axial circulation in the internal gas, which is necessary to analyze the cause of formation and influence factors for understanding better the internal flow field. There are many factors affecting the axial circulation. Different circulation drives have different effects on the flow field. In this paper, numerical computation with N-S equations is used to compute the flow field parameters and analyze the mechanism of the flow field. The influence of the temperature of the end cap on the flow field is mainly disscussed. By comparing and analyzing the streamline shape and the size of vortex region under different temperature drive, an effective method is provided for the study of axial circulation in the high-speed rotating flow field.

1 Introduction

In the rotor with high speed, the gas entering the rotor will not only make circular motion along with the rotor. In addition, due to the influence of rarefied gas state and boundary barrier, it will also form axial circulation inside the rotor, making the flow field complex and difficult to analyze. Sometimes a certain axial circulation is also needed to accelerate the flow of gas in the rotor. Therefore, it is an important subject for the study of the high-speed rotating flow field to analyze the production and influence factors of the axial circulation with an effective method [1-3].

There are many factors affecting the axial circulation. Different circulation drives have different effects on the flow field. Among these factors, the influence of temperature on the axial circulation is especially obvious. The temperature difference between the different parts of the rotor, such as the top and bottom end cap and the wall, directly affects the flow state of the internal gas. Therefore, it is of great significance to study the influence of the temperature drive on axial circulation at different parts of the rotor for analysis of the nature of the flow field.

In this paper, numerical computation with N-S equations is used to compute the flow field parameters [4] and analyze the streamline diagram and the mechanism of the flow field. The influence of the temperature of the end cap on the flow field is mainly disscussed. By comparing and analyzing the streamline shape and the size of vortex region under different temperature drive, an effective method is provided for the study of axial circulation in high-speed rotating flow field.

2 Physical model and numerical calculation

2.1 Physical model

As shown in Fig. 1. This is a physical model for a high-speed rotating flow field.

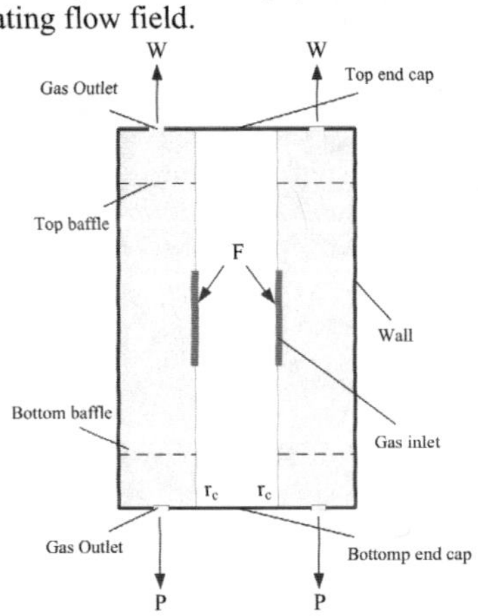

Fig. 1. Schematic diagram of physical model.

Due to the high-speed rotation of the rotor, the gas pressure in the rotor is exponentially distributed along the radial direction.. The Knudsen number is equal to 1 in the transition region, and the region is greater than 1 is the viscous flow region, which is less than 1 is the molecular flow region. It is difficult to calculate the flow field simultaneously in the transition region, the viscous flow region and the molecular flow region [5-6]. By

https://doi.org/10.1051/matecconf/201818601001

calculating, the Kn number is 1 when r_c equals 0.7 at a given speed, where the inner boundary is set. This paper only calculates the viscous flow region between the inner boundary and the wall. In the physical model, the gas inlet is set at the inner boundary, and the gas F enters from the side of the cylinder in the entrance section. A narrow annular slit is made at the top and bottom end cap to simulate the gas outlet. The gas F flows out from the two outlets forming the gas P and W. It is assumed that the gas velocity near the inlet is uniformly distributed along the axial direction and the gas velocity near the slit is uniformly distributed along the radical direction. There are top and bottom baffles in the rotor. Assuming that the top and bottom baffles are infinitely thin, annular holes are used to simulate the openings of baffles. The location and size of the annular holes are the same as those of the actual baffle holes.

2.2 Numerical calculation

The basic assumptions made in the N-S equations are:

(1) Continuous medium hypothesis;

(2) The fluid is an isotropic Newtonian fluid, and Fourier law is satisfied in the heat transfer property;

(3) The fluid satisfies the ideal gas equation of state and is in the state of thermodynamic equilibrium;

(4) Mass force is limited to gravity;

(5) The Stokes hypothesis is used for the expansion viscosity coefficient of compressible fluid, making the expansion viscosity coefficient be $\lambda = -2\mu/3$;

Under the above assumptions, the equations in the cylindrical coordinate system are expressed as follows [7]:

Continuous equation:

$$\frac{\partial\rho}{\partial t}+\frac{1}{r}\frac{\partial}{\partial r}(r\rho V_r)+\frac{1}{r}\frac{\partial}{\partial\theta}(\rho V_\theta)+\frac{\partial}{\partial z}(\rho V_z)=0 \quad (1)$$

Momentum equation:

Radial direction:

$$\rho\left(\frac{DV_r}{Dt}-\frac{V_\theta^2}{r}\right)=\rho g_r-\frac{\partial p}{\partial r}+\frac{\partial}{\partial r}\left[2\mu\frac{\partial V_r}{\partial r}-\frac{2}{3}\mu(\nabla\cdot\vec{V})\right]+\frac{1}{r}\frac{\partial}{\partial\theta}\left[\mu\left(\frac{1}{r}\frac{\partial V_r}{\partial\theta}+\frac{\partial V_\theta}{\partial r}-\frac{V_\theta}{r}\right)\right]+\frac{\partial}{\partial z}\left[\mu\left(\frac{\partial V_r}{\partial z}+\frac{\partial V_z}{\partial r}\right)\right]+\frac{2\mu}{r}\left(\frac{\partial V_r}{\partial r}-\frac{1}{r}\frac{\partial V_\theta}{\partial\theta}-\frac{V_r}{r}\right) \quad (2)$$

Angular direction:

$$\rho\left(\frac{DV_\theta}{Dt}+\frac{V_rV_\theta}{r}\right)=\rho g_\theta-\frac{1}{r}\frac{\partial p}{\partial\theta}+\frac{1}{r}\frac{\partial}{\partial\theta}\left[\frac{2\mu}{r}\frac{\partial V_\theta}{\partial\theta}-\frac{2}{3}\mu(\nabla\cdot\vec{V})\right]+\frac{\partial}{\partial z}\left[\mu\left(\frac{1}{r}\frac{\partial V_z}{\partial\theta}+\frac{\partial V_\theta}{\partial z}\right)\right]+\frac{\partial}{\partial r}\left[\mu\left(\frac{1}{r}\frac{\partial V_r}{\partial\theta}+\frac{\partial V_\theta}{\partial r}-\frac{V_\theta}{r}\right)\right]+\frac{2\mu}{r}\left(\frac{2}{r}\frac{\partial V_r}{\partial\theta}+\frac{\partial V_\theta}{\partial r}-\frac{V_\theta}{r}\right) \quad (3)$$

Axial direction:

$$\rho\frac{DV_z}{Dt}=\rho g_z-\frac{\partial p}{\partial z}+\frac{1}{r}\frac{\partial}{\partial r}\left[\mu r\left(\frac{\partial V_z}{\partial r}+\frac{\partial V_r}{\partial z}\right)\right]+\frac{1}{r}\frac{\partial}{\partial\theta}\left[\mu\left(\frac{1}{r}\frac{\partial V_z}{\partial\theta}+\frac{\partial V_\theta}{\partial z}\right)\right]+\frac{\partial}{\partial z}\left[2\mu\frac{\partial V_z}{\partial z}-\frac{2}{3}\mu(\nabla\cdot\vec{V})\right] \quad (4)$$

Energy equation:

$$\rho\frac{D(C_pT)}{Dt}=\frac{Dp}{Dt}+\Phi+\frac{1}{r}\frac{\partial}{\partial r}\left(\kappa r\frac{\partial T}{\partial r}\right)+\frac{1}{r^2}\frac{\partial}{\partial\theta}\left(\kappa\frac{\partial T}{\partial\theta}\right)+\frac{\partial}{\partial z}\left(\kappa\frac{\partial T}{\partial z}\right)+\frac{\partial Q}{\partial t} \quad (5)$$

State equation:

$$Mp=\rho RT \quad (6)$$

In the equations, the column coordinate radial, angular and axial coordinates are marked by r,θ,z. The velocity component in the direction of r,θ,z are marked by V_r,V_θ,V_z. The static pressure of gas is marked by p. The gas temperature is marked by T. The gas density is marked by ρ. The dynamic viscosity coefficient of gas is marked by μ. The specific heat at constant pressure of gas is marked by C_P. The coefficient of gas heat conductivity is marked by κ. The molecular weight of gas is marked by M. The universal gas constant is marked by R. The viscous dissipation coefficient is marked by Φ. The unit volume calorific value of heat source in the flow field is marked by Q.

In the numerical calculation process, this paper linearize the N-S equations and make them dimensionless, then use the finite difference method to discretize the dimensionless equations, so as to get the fluid parameter values of each cell [8].

3 Calculation results and discussion

3.1 Parameter setting

According to the above caculation method, the drive quantities are set which affect the axial circulation. In the physical model, the driving mechanisms that can affect the axial circulation mainly include the top and bottom mechanical drive, the wall thermal drive, the gas inlet radial, angular and axial drive. The parameters of each drive are set as shown in Table 1.

Table 1. The parameters of each drive setting.

Drive Quantity	Value
The top mechanical drive quantity	-0.11
The bottom mechanical drive quantity	-0.001
The wall thermal drive quantity	0~0.03
The gas inlet radial drive quantity	0.60
The gas inlet angular drive quantity	-0.01

The gas inlet angular drive quantity	0.16

3.2 Streamlines study

According to Tab. 1, the wall thermal drive quantity is assumed as 0.02, and fully driving streamlines are shown in Fig. 2.

From the streamlines shown in Fig. 2, it is obvious that under the common effect of mechanical drive, wall thermal drive and gas inlet drive, there are three large vortexes in the flow field. As shown in the diagram, the vortex Ⅰ is near the axial position of the gas inlet, the vortex Ⅱ is close to the wall, and the vortex Ⅲ is in the top baffle nearby. In addition, there are three small vortexes between the baffle and the end cap. The influence of the vortex produced in the flow field on the axial caculation is defferent, and further analysis is needed to draw a conclusion.

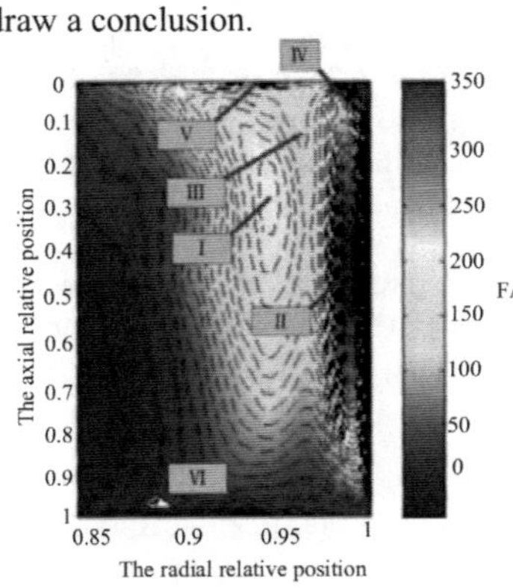

Fig. 2. Fully driving streamlines.

A streamline is a connection of the equivalent points of a flow function, the direction of whose tangent line coincides with the direction of the velocity vector. It is necessary to judge the positive and negative of the mass flow through the calculation of the flow field, and obtain the direction of the vortex. Based on that, the influence of each vortex on the axial circulation can be inferred. Fig. 3 is the section flow pattern with full drive.

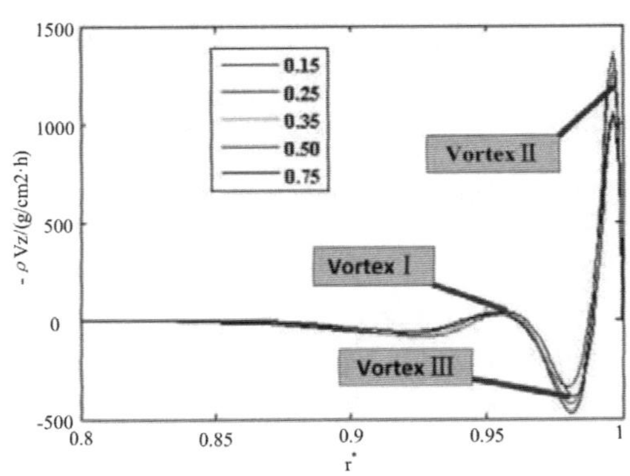

Fig. 3. The section flow pattern with full drive.

According to the positive and negative of mass flow, it can be judged that the vortex Ⅰ and Ⅱ are counter clockwise, which can be in favor of the axial circulation in the rotor. While the vortex Ⅲ is clockwise, the effect is the opposite. Based on the effect of the above vortex on the axial circulation, it is possible to further analyze the effect of temperature difference between the top and bottom end cap on vortex in the internal flow field of the rotor.

The temperature difference between the top and bottom end cap directly affects the temperature of the wall. Therefore, it is necessary to set the wall thermal drive to 0, then comparing the simulated streamlines with the streamlines under the full drive mode and observing the change of vortexes. The missing vortex is produced by the effect of the wall thermal drive.

Fig. 4 is the streamlines without the wall thermal drive. Compared with Fig. 2, Fig. 4 clearly lacks the vortex Ⅱ, while other vortex streamlines have little change. It is possible to judge that the vortex Ⅱ is mainly produced by the wall thermal drive. From Fig. 3, the existence of vortex Ⅱ can promote the axial circulation, so the wall thermal drive is needed to adjust and find the best value, so as to maximize the promotion.

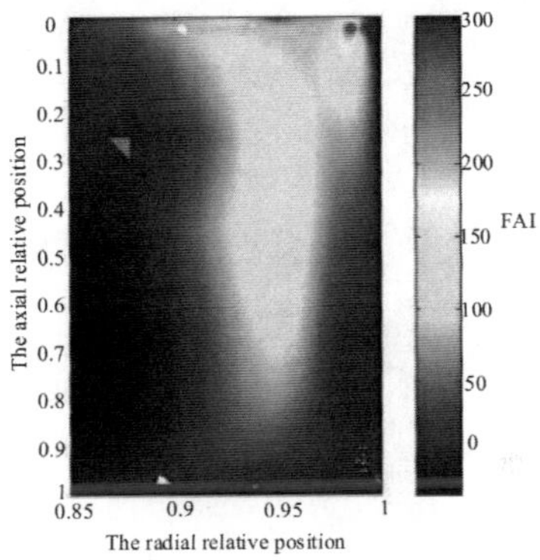

Fig. 4. The streamlines without the wall thermal drive.

3.3 Effect of different end cap temperature

Fig. 5 to Fig. 7 represent respectively the streamlines of the wall thermal drive quantity with 0.01, 0.02 and 0.03.

 https://doi.org/10.1051/matecconf/201818601001

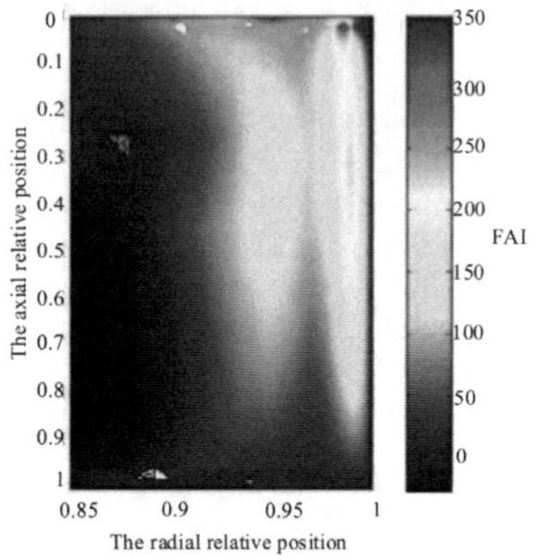

Fig. 5. The streamlines of the wall thermal drive quantity with 0.01.

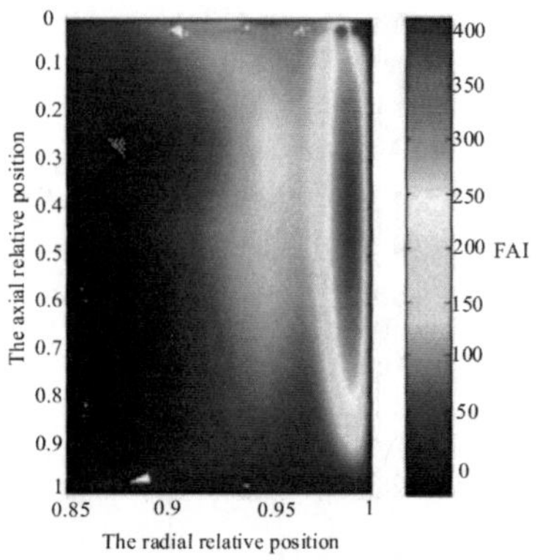

Fig. 6. The streamlines of the wall thermal drive quantity with 0.02.

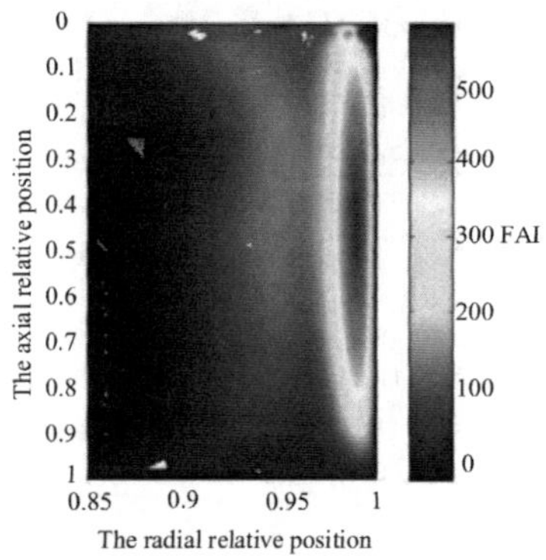

Fig. 7. The streamlines of the wall thermal drive quantity with 0.03.

From Fig. 5 to Fig. 7, when the wall thermal drive quantity is 0.01, a vortex is produced near the wall, but it is not intense. When the drive quantity increases to 0.02, the flow function value at the center of the vortex is greater than other values around it, indicating that the flow of the vortex is intensified. When the drive quantity increases to 0.03, the size and range of the vortex II remain the same. But the region of the vortex I decreases obviously and the vortex flow slows down. Meanwhile conversely, the other vortex regions remain unchanged. All of those is indicated that the increase of the wall thermal drive has an impact on certain drives, which weakens the axial circulation and increases the adverse effect. Through comparison and analysis, it is inferred that the most favorable wall thermal drive quantity to the axial circulation of the gas inside the rotor is 0.02, and the value is not better as large as possible.

4 Conclusion

In this paper, the finite difference method is used to calculate the fluid parameters under the linear distribution of the wall temperature. Through the analysis of the streamlines, when the wall thermal drive quantity is 0.02, the generation mechanism of several main vortexes in the flow field is obtained. By comparing the streamlines of different wall thermal drives, it is concluded that the wall thermal drive quantity with 0.02 is the most favorable for the axial circulation. Through this conclusion, it is possible to better control the axial circulation of the gas in the high-speed rotor, making it more conducive to the flow of the gas in the flow field, so as to better control the physical quantity of the gas at the outlet.

The temperature distribution of the wall is assumed to be linear and close to the ideal distribution in the calculation process in this paper. While the temperature of the wall in the actual high-speed rotating rotor is not linear affected by external factors, the next step will continue to calculate and analyze the flow field under different temperature distributions of the wall, especially the non-linear temperature distribution, which can be better adapt to the actual situation..

References

1. J.P.Lahargue, Soubbaramayer, J. Computer Meth. Appl. Mech. Eng **15** (1978)
2. Junni L, Xiuping L, Youzhi L, Mach. Des. Manu **2**, 7-11 (2013)
3. Mengting Jiang, Bo Wang, J. Environ. Eng **6** (2012)
4. Weixing H, Jianming L, Zeyi X, *Engineering Fluid Mechanics*, 136-140 (2008)
5. Baoguo W, Xiang L, Weiguang H, J. Areo. Pow **25**, 964-980 (2010)
6. Zhonghua L, Zhihui L, Haiyan L, Acta Aero. Sinica **31**, 282-287 (2013)
7. Jhon D.Anderson, *Computation Fluid Dynamics*, 51-55 (2007)
8. H.G.Wod, G.Sanders, J. Fluid. Mech **127** (1983)

Electromagnetic Radiation from the Geometrical Point of View

Divakov Dmitriy[1], *Malykh* Mikhail[1], and *Tiutiunnik* Anastasiia[1]

[1]Peoples' Friendship University of Russia (RUDN University), 6 Miklukho-Maklaya St, Moscow, 117198, Russian Federation

Abstract. The article describes the relationship between the solutions of Maxwell's equations which can be considered at least locally as plane waves and the curvilinear coordinates of geometrical optics. We introduce phase-ray coordinate system for any electromagnetic field if vectors **E** and **H** are orthogonal to each other and their directions do not change with time *t*, but may vary from point to point in the domain *G*.

1 Introduction

Assuming that in the domain $(x, y, z) \in G \subset \mathbb{R}^3, \quad t > 0$ with isotropic filling ε and μ two vector fields **E** and **H** are given, satisfying the homogeneous system of Maxwell's equations

$$\operatorname{rot} \mathbf{E} = -\frac{\mu}{c}\frac{\partial \mathbf{H}}{\partial t}, \quad \operatorname{rot} \mathbf{H} = \frac{\varepsilon}{c}\frac{\partial \mathbf{E}}{\partial t}. \tag{1}$$

It is well known that there are solutions of the system, which can be described by means of geometrical optics. These are, e.g., the plane waves in a homogeneous medium. Under the traditional approach it would be expected that in the general case, the laws of geometrical optics are valid only for harmonic vibrations, and only in the high-frequency limit. But yet Lüneburg [1] noted that the movement of discontinuity surfaces of electromagnetic fields obey the laws of geometrical optics exactly, and not only in the high-frequency limit. In our researches [2] we try to introduce the concepts of rays and phase fronts for solving Maxwell's equations without reference to the high-frequency limit and prove that the so-imposed rays obey the laws of geometrical optics.

2 Phase-ray coordinate system

For electromagnetic fields which can be considered at least locally as plane waves, vectors **E** and **H** are orthogonal to each other and their directions do not change with time *t*, but may vary from point to point in the domain *G*, as it occurs, for example, in the case of spherical wave. Therefore, as the first condition for the considered electromagnetic field, assume the following. Let the vectors $\mathbf{e}_2 = \pm\mathbf{E}/\|\mathbf{E}\|$ and $\mathbf{e}_3 = \pm\mathbf{H}/\|\mathbf{H}\|$ do not depend on *t* and are orthogonal to each other. The sign is chosen once in an arbitrary manner under assumption that $\mathbf{e}_2$ and $\mathbf{e}_3$ are smooth vector fields in the domain *G*. The integral curves of the vector field $\mathbf{S} = \mathbf{E} \times \mathbf{H}$ we will call the *rays* of the electromagnetic field $\{\mathbf{E}, \mathbf{H}\}$. The surface orthogonal to $\mathbf{S} = \mathbf{E} \times \mathbf{H}$ at all points, will be called the phase front of the electromagnetic field $\{\mathbf{E}, \mathbf{H}\}$.

By the assumption directions of vectors **E** and **H** do not change with time, so the vector $\mathbf{e}_1 = \pm\mathbf{S}/\|\mathbf{S}\|$ does not depend on *t* as well. The ray emanating from the point (x_0, y_0, z_0), is a trajectory of a point which motion is described by the initial problem

$$\dot{\mathbf{r}} = \mathbf{e}_1(\mathbf{r}), \quad \mathbf{r}|_{t=0} = (x_0, y_0, z_0)^T. \tag{2}$$

The existence and uniqueness of its solution follow directly from Cauchy's theorem. Therefore, no matter how complicated the considered electromagnetic field is, from any point (x_0, y_0, z_0) it is always possible to emit a ray and it will be the only one.

Subject to Jacobi theorem the necessary and sufficient condition of holonomicity of field **A** is the equality $(\operatorname{rot} \mathbf{A}, \mathbf{A}) = 0$; scalar $(\operatorname{rot} \mathbf{A}, \mathbf{A})$ is called the quantity of field **A** nonholonomicity. Therefore, if the electromagnetic field satisfies Maxwell's equations (1) and the stated above conditions, by virtue of the theorems of Cauchy and Jacobi there are well defined: 1) two-parameter family of rays, and 2) one-parameter family of phase surfaces.

If the fields **E** and **H** satisfy the Maxwell's equations (1) and the stated above conditions, they are holonomic [2]. This theorem allows one to associate the field with a curvilinear orthogonal coordinate system in which the Maxwell's equations look particularly simple. Due to the theorem of Jacobi, there are three such functions u_1, u_2 and u_3, that

$$\mathbf{S} \parallel \nabla u_1, \quad \mathbf{E} \parallel \nabla u_2, \quad \mathbf{H} \parallel \nabla u_3. \tag{3}$$

These functions define an orthogonal curvilinear coordinate system, which will be called *phase-ray coordinate system*.

Let h_1, h_2, h_3 be the Lame coefficients of the coordinate system. Using the same method [3] that is used in the theory of diffraction when introducing Borgnis functions [4], we can prove [2] that the field can be represented as

$$\mathbf{E} = \frac{E}{h_2}\mathbf{e}_2,\ \mathbf{H} = \frac{H}{h_3}\mathbf{e}_3 \tag{4}$$

where scalars E and H satisfy the system

$$\begin{cases} \frac{\partial}{\partial x_1}E + m_e\frac{\partial}{\partial t}H = 0, \\ \frac{\partial}{\partial x_1}H + m_h\frac{\partial}{\partial t}E = 0, \\ \frac{\partial}{\partial x_3}E = 0, \\ \frac{\partial}{\partial x_2}H = 0, \end{cases} \tag{5}$$

the coefficients of which m_e and m_h are expressed in terms of the characteristics of filling ε and μ in the region and the Lame coefficients by the formulas

$$m_e = \frac{\mu}{c}\frac{h_1h_2}{h_3},\ m_h = \frac{\varepsilon}{c}\frac{h_1h_3}{h_2}. \tag{6}$$

In general, the system (5) comprises only four equations and two unknown functions, so from general considerations one would expect that the condition of the solvability of this system provides some restrictions on the coefficients m_e and m_h. For the system (5) has a non-static solution it is necessary that one of its coefficients m_e and m_h can be represented as a product of the form $\phi(x_1,x_2)\psi(x_1,x_3)$ and it is enough that both coefficients m_e and m_h depend only on x_1 [2].

Sufficient conditions immediately refer us to geometrical optics. By the hypothesis in the curvilinear coordinate system under consideration the value

$$c^2m_em_h = \varepsilon\mu h_1^2 \tag{7}$$

is a positive constant, say a^2. Let the points A and B lie on a coordinate curve $x_1 = c_1$, then the optical length of the arc of any curve connecting these points is equal to the expression

$$\int_A^B \sqrt{\varepsilon\mu}ds = \int_{x_1=a}^{b}\sqrt{a^2 + \varepsilon\mu h_2^2\left(\frac{dx_2}{dx_1}\right)^2 + \varepsilon\mu h_3^2\left(\frac{dx_3}{dx_1}\right)^2}\,dx_1. \tag{8}$$

This value will be minimal, if we take as the curve the arc of coordinate curve when

$$\int_A^B \sqrt{\varepsilon\mu}ds = \int_{x_1=a}^{b}\sqrt{\varepsilon\mu}h_1dx_1. \tag{9}$$

Thus x_1-lines are rays of geometrical optics [2].

The wave propagation along the rays will be subject to the equations system (5). To investigate the type of this system let us excluded a variable H from the system (5). To this end, we differentiate the first equation of the system (5) by x_1, then we have

$$\frac{\partial^2E}{\partial x_1^2} - m_em_h\frac{\partial^2E}{\partial t^2} = \frac{\partial \ln m_e}{\partial x_1}\frac{\partial E}{\partial x_1}, \tag{10}$$

which is different from the usual hyperbolic equation by an additional term on the right side. This allows us to count on the extension of the class of problems whose solution can still be described by a system of rays and phase fronts. Equation (10) has the form

$$\frac{\partial^2u}{\partial x_1^2} - \frac{1}{c^2}\frac{\partial^2u}{\partial t^2} = a\frac{\partial u}{\partial x}, \tag{11}$$

thus, it is a classical equation of oscillations with an additional term $a\frac{\partial u}{\partial x_1}$, which in the theory of oscillations is used for modeling damping vibrations phenomena due to the resistance of the medium [5]. If we take $a = 0$ we get the usual equation of vibrations, so when choosing a certain initial conditions the instantaneous profile u will move along the x_1-axis at the velocity c. Posing $a > 0$ will lead to damping.

3 The case of constant coefficients

Using constants m_e and m_h the general solution of the system (5) can be written explicitly. Indeed, we differentiate the first equation (5) by x_1, then

$$\frac{\partial^2}{\partial x_1^2}E + m_e\frac{\partial^2}{\partial x_1\partial t}H = 0. \tag{12}$$

Using the second equation, we get

$$\frac{\partial^2}{\partial x_1^2}E - m_em_h\frac{\partial^2}{\partial t}E = 0. \tag{13}$$

Classical electrodynamics considers media with positive ε and μ, Lame coefficients can not be negative as well. Therefore $m_em_h > 0$ and this equation is the hyperbolic equation, the general solution of which is given by d'Alembert's formula

$$E = F_1(\sqrt{m_em_h}x_1 - t, x_2, x_3) + F_2(\sqrt{m_em_h}x_1 + t, x_2, x_3). \tag{14}$$

The third equation of the system indicates that these functions should not depend on x_3. Differentiating the fourth equation of the system with respect to t and using the first equation, we get $\frac{\partial^2E}{\partial x_1\partial x_2} = 0$. Therefore

$$E = F_1(\sqrt{m_em_h}x_1 - t) + F_2(\sqrt{m_em_h}x_1 + t) + F_3(x_2). \tag{15}$$

We can write out a similar formula for H. Thus, in this case, the field is a superposition of two waves traveling along the rays in opposite directions, to which the static summand is added. The presence of this summand is due to the fact that earlier we have not excluded static fields from the consideration. Obviously when considering the electromagnetic radiation it is reasonable to omit it.

The x_1-lines of the phase-ray coordinate system represent rays of geometrical optics, and the field itself is a superposition of two waves running along the rays in opposite directions:

$$\mathbf{E} = F(\sqrt{\varepsilon\mu}h_1x_1 - ct)\frac{\mathbf{e}_2}{h_2} + G(\sqrt{\varepsilon\mu}h_1x_1 + ct)\frac{\mathbf{e}_2}{h_2}. \tag{16}$$

In particular, if the field in question has a gap at $t = 0$, the gap propagates along the rays according to the laws of geometrical optics, as it should be, according to the Lüneburg [1]. Thus our approach give a true geometrical interpretation of Maxwell's equations, this is the first step on the way of the geometrization of classical electrodynamics [6-8]

4 Hilbert-Courant waves

Wishing to give a mathematical description of wave concepts, Hilbert and Courant [9] noticed that for hyperbolic equation

$$\mathfrak{L}[u]=0 \quad (17)$$

one can specify such functions θ that for any choice of the function W of one variable the expression

$$u=W(\theta(\vec{x},t)) \quad (18)$$

is the solution to the equation $\mathfrak{L}[u]$. These solutions can be interpreted as waves, the function W - as a waveform, and θ - as its phase; the form of these waves is not distorted during propagation. At the same time it can be shown that the phase necessarily satisfies the eikonal equation. This concept can be generalized to the case of distorted waves, by considering the family of solutions

$$\{u=g(\vec{x},t)W(\theta(\vec{x},t)) \quad \forall W \in C^1(\square)\}. \quad (19)$$

The function g introduced here, characterizes the distortion of the wave.

Localized wave solutions to the homogeneous scalar wave equation in free space are of three types: focus wave modes, superluminal and subluminal solutions. Each of them includes a subclass of solutions conforming to the Courant-Hilbert approach. One more class of solutions, spatiotemporally localized luminal null electromagnetic fields [10] also concern the Courant-Hilbert approach. They are transverse with respect to the local flow of energy. A class of luminal null electromagnetic waves has vortex structures on the planes transverse to the direction of propagation.

Null solutions of the Maxwell's equations were studied by Robinson in connection with his work on general relativity. A family of such null solutions of the Maxwell's equations was described, they are usually expressed in a special coordinate system.

There is a solution to the Maxwell's equations linked to a beam of light whose electric (magnetic) field lines are all closed loops with any two electric (magnetic) field lines linked to each other. Maxwell's equations preserve for such solutions the topological structure of the field lines. In [11] it was shown that such an evolution of the field is due to the condition that the electric and magnetic fields are everywhere orthogonal. For the wave equation one can find a variety of such waves [12].

As a part of our method Hilbert-Courant approach can be transferred to the vector case.

Non-permanent solution of (5)

$$\begin{aligned} E&=E(\theta(x_1,t)),\\ H&=H(\theta(x_1,t)) \end{aligned} \quad (20)$$

will be called the wave propagating along the ray without distortion, the function $\theta(x_1,t)$ will be called the phase of the wave.

The phase of any wave satisfies the eikonal equation

$$\left(\frac{\partial\theta}{\partial x_1}\right)^2-\frac{\varepsilon\mu h_1^2}{c^2}\left(\frac{\partial\theta}{\partial t}\right)^2=0. \quad (21)$$

If the product $m_e m_h$ depends only on x_1, and the ratio m_e/m_h is a constant, then any non-permanent solution of the eikonal equation can be in the form of a wave

$$\begin{aligned} E&=W(\theta(x_1,t)),\\ H&=\pm\sqrt{\frac{m_h}{m_e}}W(\theta(x_1,t)) \quad \forall W\in C^1(\square), \end{aligned} \quad (22)$$

propagating along the ray without distortion.

The transition from auxiliary functions E, H to the electromagnetic field by the formulas (4) introduces additional factors which depend on the Lame coefficients. Therefore, the solution of the eikonal equation can be in the form of a wave propagating along the ray with a certain distortion of the form due to curvature of rays.

5 Monochromatic field at high frequencies

Assuming that the solution (10) depends on time as $E=u(x_1)\sin\omega t$, and assuming for brevity $x_1=x$, we obtain

$$\frac{d^2u}{dx^2}-\frac{1}{m_e}\frac{\partial m_e}{\partial x}\frac{du}{dx^2}+\omega^2 m_e m_h u=0. \quad (23)$$

In the high-frequency limit $\omega\to\infty$ one can try to find a solution by the JWKB method. Let us consider a solution as a series

$$u=\exp\left(\frac{1}{\lambda}\int_0^x dx(u_0+\lambda u_1+\ldots)\right) \quad (24)$$

in powers of a small parameter $\lambda=1/\omega$. Thus, at high frequencies, any monochromatic solution of the auxiliary equation (10) is a superposition of scalar waves of the form

$$\sqrt[4]{\frac{m_e}{m_h}}\sin\left(t\pm i\omega\int\sqrt{m_e m_h}\,dx+\text{const}\right), \quad (25)$$

traveling along the ray in two directions. The dependence of $\sqrt[4]{m_e/m_h}$ on x_1 characterizes the change in the amplitude of these waves along the ray.

Let us verify the formula (25) by numerical experiments. Figures 1,2 shows the solutions of the equation (23) with $\omega=10$ for two types of filling $m_e=1/(1+x^2)$, $m_h=1$ and $m_e=1+x$, $m_h=1$, found by the Runge-Kutta method, the initial data are chosen in a way to achieve the best coincidence with the asymptote $x=0$. The graphs clearly show that the formula (25) gives the right course of solving the equation.

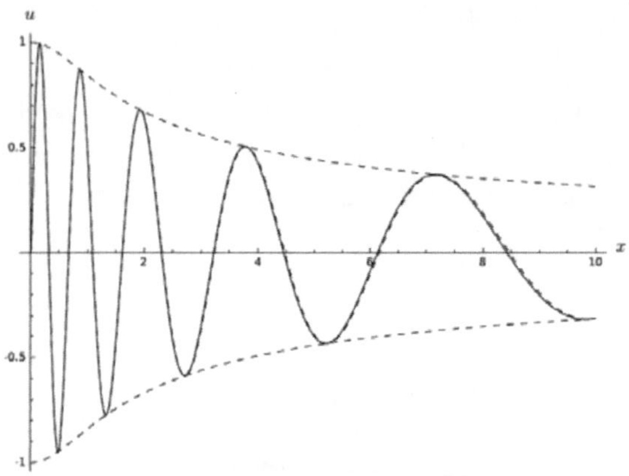

Fig.1. Solutions of the equation (23) and his asymptotics at $\omega = 10$ for $m_e = 1/(1+x^2)$, $m_h = 1$.

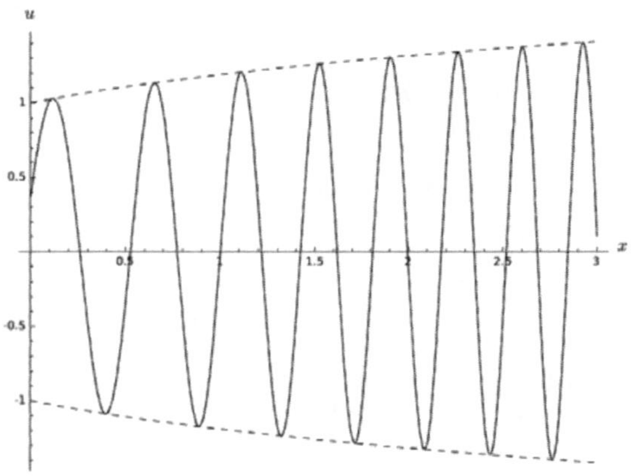

Fig.2. Solutions of the equation (23) and his asymptotics at $\omega = 10$ for $m_e = 1 + x$, $m_h = 1$.

6 Conclusion

On the assumption that in the region G there is a homogeneous solution of Maxwell's equations, satisfying additional conditions of holonomicity of the Poynting vector field we build locally-orthogonal coordinate system (called the phase-ray system) having a clear geometric-optical interpretation. It is shown that under certain non-burdensome (sufficient) conditions on variable real coefficients of medium filling in the region (on the permittivity and magnetic susceptibility), a phase-ray coordinate system that describes the laws of geometrical optics in this region, can be constructed, and there exist solutions of Maxwell's equations, which take in this coordinate system the simplest (canonical) form.

The publication has been prepared with the support of the "RUDN University Program 5-100".

References

1. R. Lüneburg, *Mathematical theory of optics. Providence* (Rhode Island: Brown University, 1944)
2. M. D. Malykh, N. E. Nikolaev, L. A. Sevastianov, The Geometrical Description of Electromagnetic Radiation. Journal of Electromagnetic Waves and Applications **30(15),** 2055-2066 (2017)
3. M. D. Malykh, On solutions of the Maxwell's equations from the viewpoint of geometrical optics. Bull PFUR, ser MIPh. **4**, 65-72 (2014)
4. K. Zhang, D. Li, *Electromagnetic theory for microwaves and optoelectronics* (Berlin: Springer, 2007)
5. W. Yang, *Reflection Seismology* (Elsevier, 2014)
6. D.S. Kulyabov, A.V. Korolkova, L.A. Sevastianov, Spinor representation of Maxwell's equations. Journal of Physics: Conference Series. **788(1),** 012025 (2017)
7. D.S. Kulyabov, A.V. Korolkova, L.A. Sevastianov, M.N. Gevorkyan, A.V. Demidova, Geometrization of Maxwell's equations in the construction of optical devices. Proceedings of SPIE - The International Society for Optical Engineering. **10337**, 103370K (2017)
8. D.S. Kulyabov, A.V. Korolkova, L.A. Sevastianov, E.G. Eferina, T.R. Velieva, A geometric approach to the Lagrangian and Hamiltonian formalism of electrodynamics. Proceedings of SPIE - The International Society for Optical Engineering. **10337**, 103370M (2017)
9. R. Courant, D. Hilbert, *Methoden der mathematischen Physik* (Springer Verlag, 1968)
10. I.M. Besieris, A.M. Shaarawi, Spatiotemporally localized null electromagnetic waves I. Luminal. Progress In Electromagnetics Research B **8**, 1-28 (2008)
11. W.T.M. Irvine, Linked and knotted beams of light, conservation of helicity and the flow of null electromagnetic fields. J. Phys. A: Math. Theor. **43**, 815-820 (2010)
12. I. M. Besieris, A. Shaarawi, Three classes of Courant-Hilbert progressive solutions to the scalar wave equation. Journal of electromagnetic waves and applications **16(8)**, 1047-1060 (2002)

A New Explanation for the Color Variety of Photons

Gh. Saleh[1], *M. J.* Faraji[1], *R.* Alizadeh[1], and *A.* Dalili[1]

[1]Saleh Research Centre, Shiraz, Fars, Iran

Abstract. This new explanation is based on *Wave-Particle Duality* and *Newtonian Laws* and represents a unique definition of a three-dimensional motion for the photon, whose dual behavior is partly explained by the *double-slit experiment* of Thomas Young, who represents the photon's motion as a wave, and by the *Photoelectric effect*, in which the photon is considered as a particle. However, for scientists, the photon's true motion is unclear. In this article, we define a new type of motion for photons to solve both this ambiguity and the difficulty of presenting a three-dimensional trajectory for the photon's motion, and present a new formula to calculate its energy. In addition, because we believe in the helical motion of photons, where r is the gyroradius, we believe that their color is an effect of the order of magnitude of *r*. We present real examples that prove our energy formula.

1 Introduction

"It seems as though we must use sometimes the one theory and sometimes the other, while at times we may use either. We are faced with a new kind of difficulty" When Einstein said [1] this, he was curious to know more about the photon's behavior.

Actually, when a photon is emitted by an excited electron, it has two different motions: a rotation along and around a straight axis at light speed (v), and a three-dimensional trajectory combining both rectilinear and orbital trajectories. However, a side view of this movement still shows a sine wave (Fig. 1).

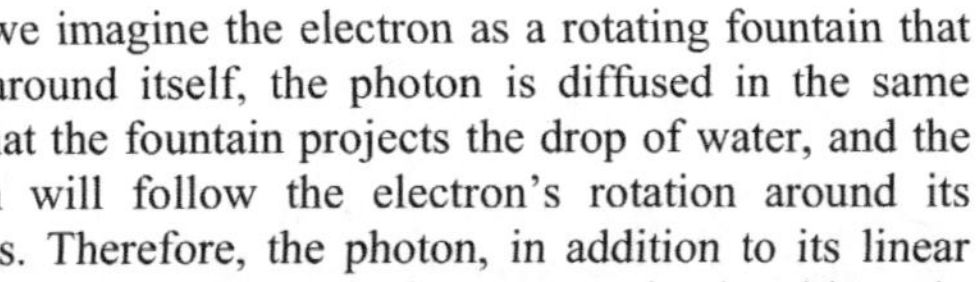

Fig. 1. Side view of a photon's motion appears as a sine wave.

If we imagine the electron as a rotating fountain that turns around itself, the photon is diffused in the same way that the fountain projects the drop of water, and the photon will follow the electron's rotation around its nucleus. Therefore, the photon, in addition to its linear trajectory, also turns around an assumed axis, ultimately creating a three-dimensional trajectory. To approach an understanding of this, we consider the example of a released spring. That is to say, the photon's trajectory is a result of the combination of two motions: the rectilinear one and a perpendicular circulation around the same axis (z).

Actually, due to the intensity of the energy released during the emission, the photon rotates both around its own axis (like the Earth's rotation), and around an external point (similar to the annual motion of the Earth) while continuing its rectilinear trajectory. Therefore, one turn of the photon around the external point is equal to the wavelength of a sinusoidal wave. Thus, the number of turns of the photon in a second is called its "Frequency" or "ϑ". Therefore, the energy of photons consists of two separate parts: the first is rotational energy, which depends on its distance from the center (r). The second depends on the number of turns per unit of time.

We believe that the photon has constant mass (m); therefore, its kinetic energy with the consideration of classic mechanics [2] is equal to:

$$E_{rot} = {}^{1}/_{2}\, I\omega^2 \quad (1)$$

where (I) is the moment of inertia, and (ω) is angular speed.

Moreover, because a photon has a constant and small mass (m), it also maintains a constant distance from the center of rotation (r) [2]; therefore,

$$I = mr^2 \quad (2)$$

As a result, its rotational kinetic energy is equal to:

$$E_{rot} = {}^{1}/_{2}\, mr^2\omega^2 \quad (3)$$

In this paper, we denote E_{rot} as the **"inner part of the kinetic energy"**

The energy of the photon also depends on the number of turns in a unit of time, or its frequency, which is expressed by the Planck formula [3]:

$$E_1 = h\vartheta \quad (4)$$

where h is Planck's constant and ϑ is the frequency.

In this paper, we denote E_1 as **"the wave part of the kinetic energy"** (Fig. 2).

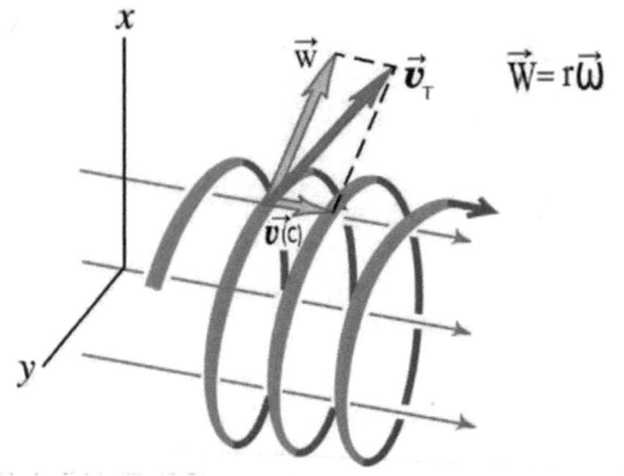

Fig. 2. Diagram of 3D motion of photon.

As a result, the total kinetic energy of a photon is equal to the sum of both the wave and the inner parts:

$$E_{Tot} = E_1 + E_{rot} = h\vartheta + {}^1/_2\, mr^2\omega^2 \quad (5)$$

However, when we consider the photon as a particle it will eventually displaced along one axis (for example Z) with a velocity of (V); therefore, its kinetic energy is equal to [4]:

$$\mathrm{E} = \int F.dz \quad (6)$$

$$F = \frac{dP}{dt} = \frac{d(mv)}{dt} \quad (7)$$

$$E = \int \frac{d(mv)}{dt} dz = \int d(mv) \frac{dz}{dt} = m \int v dv \quad (8)$$

$$E = {}^1/_2\, mv^2 \quad (9)$$

Consequently,

$$E = {}^1/_2\, mv^2 = h\vartheta + {}^1/_2\, mr^2\omega^2 \quad (10)$$

$$h\vartheta = {}^1/_2\, mv^2 - {}^1/_2\, mr^2\omega^2 \quad (11)$$

$$h\vartheta = {}^1/_2\, m(v^2 - r^2\omega^2) \quad (12)$$

where m is the photon's mass, v its velocity in a straight line, r its gyroradius, h the Planck constant, ϑ its frequency, and ω its angular velocity. In this formula, v is the velocity of the photon along the imaginary axis and equals the speed of light C, and ω has a constant value in the consumable part.

The ${}^1/_2\, mr^2\omega^2$ represents the inner part of the kinetic energy and h ϑ represents the wave part. Because m, h, ω and v are constant in one area, Formula (12) represents that the energy of photons is dependent on the frequency (ϑ) and gyroradius (r). However, the total kinetic energy is constant. An increase in frequency will decrease the gyroradius, and vice versa.

Imagine that a photon is traversing a vacuum with a constant velocity "V = C". In this condition, Equation (10) is

$$E = {}^1/_2\, mc^2 \quad (13)$$

$${}^1/_2\, mc^2 = h\vartheta + {}^1/_2\, mr^2\omega^2 \quad (14)$$

$${}^1/_2\, mc^2 - {}^1/_2\, mr^2\omega^2 = h\vartheta \quad (15)$$

In addition, because C and m are constant, the photon's energy is constant. If this photon has a frequency such as ϑ_1 its gyroradius will be r_1. If its frequency equals ϑ_2, its gyroradius will be r_2, and ${}^1/_2\, mc^2$ always has a constant value.

2 The Colorfulness of Photons

The visible spectrum is a concept that has attracted much attention. We believe that all photons are the same; they have a constant mass, and that their spectral distinction is due to the gyroradius. Changes in the gyroradius induce color changes. In fact, an increase in the gyroradius reciprocally means the lengthening of the wavelength. For example, the long wavelength (λ) of the red spectrum is a result of a high value of the gyroradius r, while the blue spectrum is the opposite (Fig. 3).

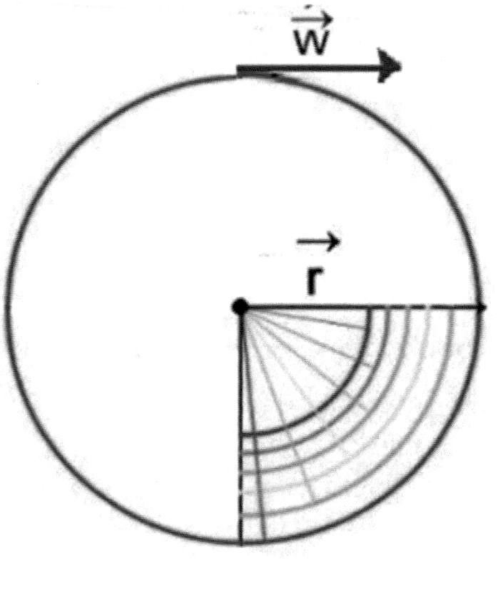

Fig. 3. Diagram of gyroradius.

We therefore consider a photon as a particle with a constant mass that generates different spectra following the variations in its gyroradius, and, indeed, that color is an effect of the order of magnitude of r.

3 Domain of ϑ, λ, r

We assume that the mass of the photon is $m = 1.637 \times 10^{-35}$ kg and that the approximate value of its angular velocity is equal to $= 1.67 \times 10^{15}\ {}^{rad}/_{s}$. The gyroradius can be calculated for three spectra: violet λ=380 n.m., yellow λ=580 n.m. and red λ=750 n.m., using equation (15):

$$\frac{1}{2}mc^2 - \frac{1}{2}mr^2\omega^2 = h\vartheta$$

$$mc^2 - mr^2\omega^2 = \frac{2hc}{\lambda} \Rightarrow mr^2\omega^2 = mc^2 - \frac{2hc}{\lambda} \Rightarrow r^2 = \frac{c^2}{\omega^2} - \frac{2hc}{\lambda m\omega^2}$$

$$\Rightarrow r = \sqrt{\frac{c^2}{\omega^2} - \frac{2hc}{\lambda m\omega^2}}$$

$$\lambda_1 = 380nm \Rightarrow \vartheta = 780\times 10^{12} Hz \Rightarrow r = 97.9nm$$

$$\lambda_2 = 580nm \Rightarrow \vartheta = 517\times 10^{12} Hz \Rightarrow r = 131.2nm :$$

$$\lambda_3 = 750nm \Rightarrow \vartheta = 400\times 10^{12} Hz \Rightarrow r = 143.6nm$$

Red:

$$\vartheta = 400 \times 10^{12}{}_{Hz} \quad (16)$$

$$r = 143.6 \times 10^{-9}{}_{m} \quad (17)$$

$$\frac{1}{2}mc^2 - \frac{1}{2}mr^2\omega^2 = \frac{1}{2} \times 1.637 \times 10^{-35} \times (2.998 \times 10^8)^2 - \frac{1}{2} \times 1.637 \times 10^{-35} \times (143.6 \times 10^{-9})^2 \times (1.67 \times 10^{15})^2 = 2.65 \times 10^{-19}j$$

$$h\vartheta = 6.626 \times 10^{-34} \times 400 \times 10^{12} = 2.65 \times 10^{-19}j$$

Yellow:

$$\vartheta = 517 \times 10^{12}{}_{Hz} \quad (18)$$

$$r = 131.2 \times 10^{-9}{}_{m} \quad (19)$$

$$\frac{1}{2}mc^2 - \frac{1}{2}mr^2\omega^2 = \frac{1}{2} \times 1.637 \times 10^{-37} \times (2.998 \times 10^8)^2 - \frac{1}{2} \times 1.637 \times 10^{-37} \times (131.2 \times 10^{-9})^2 \times (1.67 \times 10^{15})^2 = 3.42 \times 10^{-19}j$$

$$h\vartheta = 6.626 \times 10^{-34} \times 517 \times 10^{12} = 3.42 \times 10^{-19}j$$

Violet:

$$\vartheta = 780 \times 10^{12}{}_{Hz} \quad (20)$$

$$r = 97.9 \times 10^{-9}{}_{m} \quad (21)$$

$$\frac{1}{2}mc^2 - \frac{1}{2}mr^2\omega^2 = \frac{1}{2} \times 1.637 \times 10^{-37} \times (2.998 \times 10^8)^2 - \frac{1}{2} \times 1.637 \times 10^{-37} \times (97.9 \times 10^{-9})^2 \times (1.67 \times 10^{15})^2 = 5.17 \times 10^{-19}j$$

$$h\vartheta = 6.626 \times 10^{-34} \times 780 \times 10^{12} = 5.17 \times 10^{-19}j$$

Given the main formula of the Saleh Theory (15) the gyroradius change diagram (Table 1) and domain for ϑ, λ, and r is defined as follow:

$$10^{15}{}_{Hz} \geq \vartheta \geq 0_{Hz} \quad (22)$$

$$270_{n.m} \leq \lambda \leq \infty \quad (23)$$

$$0_{n.m} \leq r \leq 180_{n.m} \quad (24)$$

Table 1. Gyroradius change diagram according to Saleh Theory.

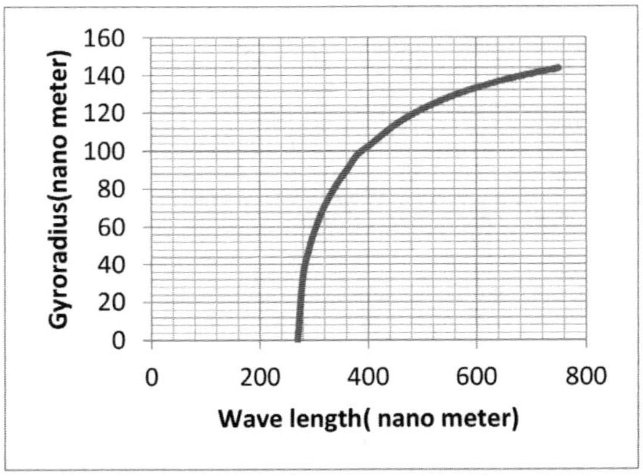

4 Example

For example, when the frequency of a photon is $\vartheta_1 = 400_{THz}$, its gyroradius will be $r_1 = 143.6_{n.m}$ and the spectrum appears red. When a photon's frequency is $\vartheta_2 = 780_{THz}$, its radius will be $r_2 = 97.9_{n.m}$ and its spectrum will appear violet.

Actually, because the red photon has a different inner kinetic energy from the violet one, our eyes interpret the differences in internal energy such that we perceive them as different colors. This is so important that in our daily life that all warning lights are red, not violet. Because the red spectrum has more inner kinetic energy than the violet one, a red spectrum with the same intensity is visible from a greater distance. If we consider only the wave part of the kinetic energy, according to the Planck and Einstein energy formulae ($E = mc^2 = h\vartheta$) [3, 5] the red spectrum has less energy than violet one. Moreover, how can a spectrum with low energy (red spectrum) traverse a greater distance than a violet one with high energy?

This is a Contradiction!

Saleh Theory states that the red spectrum has the same total kinetic energy as the violet and that they have only one important difference; the inner part of the kinetic energy makes the red spectrum more visible than the violet one from a further distance (Fig. 4).

For a better understanding, we explain another contradiction: Everyone knows that the red spectrum is warm and the blue one is cold [6]. In addition, in physics, being warm means having more energy [7]. This sense and understanding are actually the realization of the internal energy of these spectra. According to the energy formula of Saleh Theory as explained in (8), the red spectrum has a larger gyroradius than does the blue one; thus, it has a higher internal energy. Although their total energies are equal, we receive more internal energy from the red spectrum than from the blue one and we associate the red spectrum with warmth and the blue one with coldness.

Now look at the Planck formula ($E = h\vartheta$) [3] on one hand, and combine it with the Einstein formula $E = mc^2$ [5] on the other hand. The red spectrum has a lower frequency than the blue one, and it also has a lower mass; therefore, in this condition, the red spectrum, which all of us believe is warm, has lower energy and lower mass ($E = h\vartheta = mc^2$) [3, 5].

This is an Obvious Contradiction!

The third reason for the energy formula's proof and the three-dimensional rotational motion of photon:
The frequency formula of the *Doppler Effect in light* is [8]:

$$\vartheta_1 = \vartheta_2\sqrt{1 - \frac{v^2}{c^2}} \quad (25)$$

that is, the result of Einstein's relativity and the Planck formula. Moreover surely

$$\vartheta_1 \neq \vartheta_2 \quad (26)$$

If we multiply both sides by Planck's constant (h) we have:

$$h\vartheta_1 \neq h\vartheta_2 \quad \Rightarrow E_1 \neq E_2 \quad (27)$$

Therefore, the energy sent from the source is different from the energy that we receive.
Now we focus on *redshift;* when the source producing the blue spectrum is moving away from us, we perceive it as red [8]. Therefore, we have:

$$\vartheta_b > \vartheta_r \Rightarrow h\vartheta_b > h\vartheta_r \Rightarrow E_b > E_r \quad (28)$$

where E_b is the energy that is emitted, and E_r is the energy that we receive, ϑ is the frequency, h is Planck's constant and the b & r indexes are reagents of the blue and red spectra.

The energies of the red spectra are lower than those of the blue ones [3, 5]. On the other hand, because Einstein's formula is $E = mc^2$ [5],

$$m_b c^2 > m_r c^2 \Rightarrow m_b > m_r \quad (29)$$

The red spectrum that we perceive has a lower mass than blue one that is emitted from the source.

$$E_b > E_r \Rightarrow E_b - E_r > 0 \quad (30)$$

$$m_b > m_r \Rightarrow m_b - m_r > 0 \quad (31)$$

Therefore, amounts of mass and energy are destroyed at the same time, which violates the principle of conservation of energy and mass [9].
However, in Saleh Theory, because $E = h\vartheta + {}^1/_2\, mr^2\omega^2$ explained in (5), the total energy of the photon is always constant, and the spectral color change from blue to red is a conversion of energy from the wave part ($h\,\vartheta$) to the inner part (${}^1/_2\, mr^2\omega^2$);

$$E_b = E_r \quad (32)$$

$$h\vartheta_b + {}^1/_2\, mr_b^2\omega^2 = h\vartheta_r + {}^1/_2\, mr_r^2\omega^2 \quad (33)$$

where m, h and ω are constant values.
It is true that $\vartheta_b > \vartheta_r$ but because $r_b < r_r$, always, we have:

$$E_b = E_r \quad (34)$$

5 Conclusions

Guided by this theory, we can establish a complete three-dimensional vision of the photon's motion, which is justified by Wave–Particle Duality and Newtonian Laws. This vision explains how the variety of light spectra are due to different gyroradius of photon during its trajectory around an imaginary axis (Fig. 4).

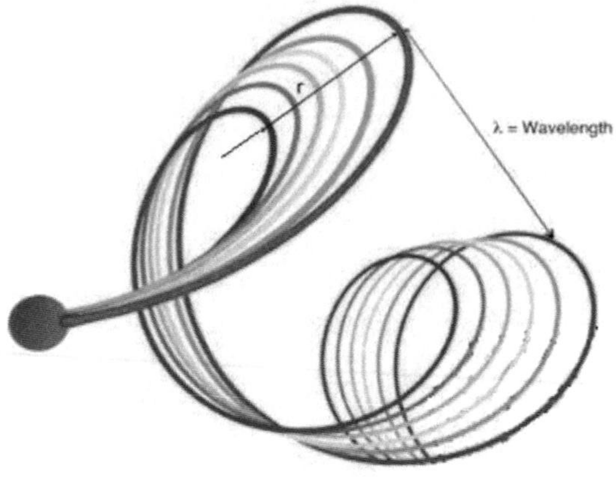

Fig. 4. Diagram of different gyroradius during a photon's trajectory around an imaginary axis.

We assume that the mass of the photon is $m = 1.637 \times 10^{-35}$ kg and that the approximate value of its angular velocity is equal to $= 1.67 \times 10^{15}\ {}^{rad}/_{s}$. The gyroradius can be calculated for three spectra;

Given the mass value of a photon and ω, the gyroradius for the visible spectrum should be between 100 nm and 150 nm (Fig.5).

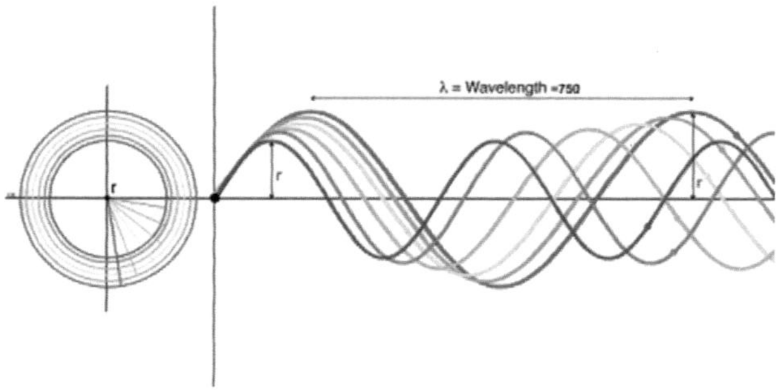

Fig. 5. Diagram of the color Variety of Photons.

References

1. D. Harrison, *Complementarity and the Copenhagen Interpretation of Quantum Mechanics*, *UPSCALE*, Dept. of Physics, U. of Toronto, (2002)
2. G.R. Fowles, Mechanics of Rigid Bodies: Planar Motion, *Analytical Mechanics, 7th edition,* (London, U. K: THOMSON, 1987)
3. S. Gasiorowicz, The limits of Classical Physics, *Quantum Physics*, (New York, NY, USA: JOHN WILEY&SONS, 1974)
4. G.R. Fowles, Newtonian Mechanics: Rectilinear Motion of a Particle, *Analytical Mechanics, 7th edition,* (London, U. K: THOMSON, 1987)

https://doi.org/10.1051/matecconf/201818601003

5. R.T. Weidner, R. Sells, Relativistic Dynamics: Momentum and Energy, *Elementary Modern Physics, 4th edition*, (New York, NY, USA: Allyn and Bacon, 1976)
6. D. Klinghardt, The Neurophysiology of light, Journal of Optometric Phototherapy, 35-40, (2003)
7. M.W. Zemansky, R.H. Dittman, Heat and the First Law of Thermodynamics, *Heat and Thermodynamics, 7th edition*, (New York, NY, USA: McGraw-Hill, 1997)
8. D. Haliday, R. Resnick, J. Walker, wave II, *Fundamentals of physics, 5th edition,* (New York, NY, USA: JOHN WILEY&SONS, 1997)
9. D. Haliday, R. Resnick, J. Walker, Potential Energy and Conservation of Energy, *Fundamentals of physics, 5th edition,* (New York, NY, USA: JOHN WILEY&SONS, 1997).

Solving Bateman Equation for Xenon Transient Analysis Using Numerical Methods

Zechuan Ding

[1]Illume Research, 4205 Xintianshiji Business Center, 125 Shixia Road, Shenzhen, China

Abstract. After a nuclear reactor is shutdown, xenon-135, an isotope with a very high thermal neutron absorption cross-section, will build up and reduce the reactivity considerably for a while. This is known as poisoning. However, the concentration xenon-135 would gradually decrease through decaying or absorbing neutron, making it necessary to suppress the reactivity in order to prevent the reactor to go critical or supercritical. Thus, it is important to predict the relationship between xenon poisoning and time after the reactor is shutdown to ensure the safety of the reactor. This paper reports the research on the prediction of xenon poisoning in a hypothetical nuclear reactor after it is shut down. In order to make the prediction, the Bateman equations of xenon (Xe) and iodine (I), which is of the form of an Ordinary Differential Equation (ODE) system, need to be solved. Two different methods, the fourth-order Runge-Kutta method and the matrix exponential method, were applied to solve the ODE system with MATLAB codes. The accuracies and computational efficiencies of the two method is also studied and compared.

1 Introduction

Ensuring safety is one of the most important goals, if not the most important goal, in today's nuclear energy industry. In most of the operating fleet of reactors, the nuclear fuel, in particular uranium-235, absorbs the thermal neutrons to maintain chain reaction to provide fission energy. However, there are many other isotopes present in the core that compete with uranium-235 for thermal neutrons, of which the isotopes with notably large cross-sections are referred to as reactivity poisons. Poisons can affect the reactivity management in many ways. Some poison materials are used in reactors for reactivity control or fuel management, whereas some fission products poisons, for example, can cause trouble to reactivity control and inherent safety of reactors.

Among all the fission products, xenon-135 has the most substantial effect on the reactivity due to its large absorption cross-section (one group absorption cross-section of 2.75 Mbarns in a standard PWR) [1]. This phenomenon is called xenon poisoning. After the shutdown of the reactor, the flux of thermal neutron will drop to almost zero. Consequently, the Xe-135 stops being produced from fission. Therefore, it is only produced through the decay of another nuclide iodine-135. However, similarly, Xe-135 does not vanish due to the absorption of thermal neutron, and it only vanishes through its β decay. Because Xe-135 has a longer half-life (9.2 hours) than I-135 (6.7 hours), xenon decays at a lower rate than iodine does. As a result, a large amount of Xe-135 will build up in the reactor, and it absorbs a significant number of neutrons. The population of xenon will keep increasing until it reaches a peak and then decline. During this process, the reactivity of the reactor will drop at first and rise again, forming a period of time which the reactor cannot be restarted due to the very low reactivity, called the iodine pit.

More importantly, the decay of xenon after the xenon peak shutdown could release large amount of reactivity into the core. Even after reactor shutdown and the control rods are fully inserted, if the reactivity increase from the decay of xenon is too high, the reactor may reach criticality or supercriticality, threatening the safety of the reactor. When designing a reactor, the control rod system must be designed such that the shutdown margin (the reactivity difference between the core at hot full power and the core at cold zero power when xenon is fully decayed after shutdown, while the control rod with the largest worth is stuck out of the core) is maintained sufficiently positive at all times. Therefore, the precise prediction of the xenon poisoning is crucial in reactor operation and the design of new reactors. Conventional tools used to model xenon transient behaviours include stochastic or deterministic lattice physics codes such as Serpent or WIMS or nodal diffusion codes such as DYN3D and Panther. These codes perform flux calculation for each time step of the transient process and relatively accurately predicts spatial xenon distribution at each step, therefore predicting its transient behavior. This paper proposes to simplify the process of xenon transient calculation by numerically solve the one group Bateman equation using MATLAB for the iodine (I) – xenon (Xe) decay chain using homogeneous reactor properties. By doing so, quicker and reasonably accurate results can be achieved which will be beneficial for preliminary reactor designs studies.

This paper starts by introducing the Bateman equation, which describes the rate of the population change of a nuclide in a reactor. The Bateman equations for different isotopes form of a system of ordinary differential equations. This study tries to solve the system of Bateman equations governing the iodine-xenon decay chain to calculate the population change of nuclide xenon-135 after reactor shutdown, which could be used to calculate the xenon population. Two methods, the fourth-order Runge-Kutta (RK4) method and the matrix exponential method, are applied to solve the equation through codes. The performances of both methods are studied in terms of accuracy and efficiency. The fourth order Runge-Kutta method is found better in terms of calculation speed; however, the matrix exponential generates results with much higher accuracy. The two methods are then used to produce 3D plot of the "xenon poisoning surface" at different thermal neutron flux levels.

2 Bateman equation for I-Xe decay chain

Bateman equation is a mathematical model that describes the abundance of a nuclide in the decay chain of isotopes. In order to calculate the population of a particular isotope, one would solve the Bateman equation governing the decay chain that includes the very isotope of interest. The following is the decay chain of the fission product tellurium-135 that involves the production and decay of Xe-135 [1]:

$$^{135}_{52}\mathrm{Te} \rightarrow {}^{135}_{53}\mathrm{I} \rightarrow {}^{135}_{54}\mathrm{Xe} \rightarrow {}^{135}_{55}\mathrm{Cs} \rightarrow {}^{135}_{56}\mathrm{Ba}$$

Let $I(t)$ and $X(t)$ be the population of iodine-135 and xenon-135 in the reactor. The equations for the I-135 and Xe-135 populations in a reactor, I and X, can then be written as:

$$\frac{d}{dt}I(t) = \lambda_T T(t) - \lambda_I I(t) \#(1)$$

$$\frac{d}{dt}X(t) = \lambda_I I(t) - \lambda_X X(t)\#(2)$$

where λ_T is the decay constant for tellurium-135,
λ_I is the decay constant for I-135, and
λ_X is the decay constant for Xe-135.

The half-life of tellurium (Te) is very short, which is roughly 11 seconds [1]. As a result, it can be assumed that I-135 is the primary fission product. Therefore, the first decay term, $\lambda_T T$ is ignored in equation (1).

Since the primary source of iodine is the fission, the rate of fission production of iodine is introduced to the equation. The equation can be then written as:

$$\frac{d}{dt}I(t) = \gamma_I \Sigma_f \phi - \lambda_I I(t)\#(3)$$

where γ_I is the fission yield of I-135, and
$\Sigma_f \phi$ is the fission rate.

The microscopic absorption cross-section of I-135 is so small that the loss rate of I-135 through absorbing neutron is very low compared to the decay rate and therefore can be ignored [1].

Besides the decay of iodine-135, fission can also contribute to the production of Xe-135. Therefore, the fission rate of Xe-135 is introduced to the Bateman equation. Besides the decay of Xe-135, the absorption of thermal neutron can also lead to the loss of Xe-135. Therefore, the full expansion of Bateman equation of xenon is obtained, which can be written as [1]:

$$\frac{d}{dt}X(t) = \gamma_X \Sigma_f \phi + \lambda_I I(t) - \lambda_X X(t) - \sigma_{aX} X(t)\phi\#(4)$$

where γ_X is the fission yield of Xe-135,
σ_{aX} is the microscopic absorption cross-section of Xe-135, and
ϕ is the neutron flux.

The Bateman equations of iodine and xenon in equation (3) and (4) form a system of initial value ordinary differential equations. By solving the ODE system, the population of xenon and iodine after the shutdown of the reactor can be obtained as a function of time. When the reactor is shut down, the neutron flux ϕ is assumed to be zero instantaneously. Let I_0 and X_0 be the population of iodine and xenon at the time of shutdown. For the reactors that have been working long enough to reach equilibrium, I_0 and X_0 can be calculated through the equations since the rate of change of the iodine and xenon populations, $\frac{d}{dt}I(t)$ and $\frac{d}{dt}X(t)$ are all zero.

By applying integrating factor technique, representations of the population of iodine and xenon can be obtained:

$$I(t) = \frac{\gamma_I \Sigma_f \phi}{\lambda_I} e^{-\lambda_I t}\#(5)$$

$$X(t) = \Sigma_f \phi \left[\frac{(\gamma_I + \gamma_X)}{\lambda_X + \sigma_{aX}\phi} e^{-\lambda_X t} + \frac{\gamma_I}{\lambda_I - \lambda_X}\left(e^{-\lambda_X t} - e^{-\lambda_I t}\right)\right]\#(6)$$

3 Methods to solve Bateman equation

3.1 Runge-Kutta method

The Bateman equations of iodine and xenon form a system of initial value ordinary differential equations. Many methods can be used to solve the initial-value ordinary differential equations with various orders, such as Euler method, midpoint method, etc. In this paper, the Fourth-order Runge-Kutta method and the matrix exponential method were selected to numerically solve the Bateman equation.

The Runge-Kutta method was developed by C. Runge and M. W. Kutta. It uses combinations of explicit and implicit iterative methods in temporal discretization to approximate solutions of ordinary differential equations. The well know Euler Method is also viewed as one of Runge-Kutta schemes [2].

Consider a general initial value ODE:

$$y' = f(t, y), \qquad y(t_0) = y_0 \#(7)$$

In the equation, y is a function of time t, and the derivative of y is a function of t and y. The value of y is y_0 at the initial time t_0.

The general form of each step in the solution of the fourth-order Runge-Kutta method is [3]:

$$y_{n+1} = y + \frac{1}{6}(\Delta y_1 + 2\Delta y_2 + 2\Delta y_3 + \Delta y_4)\#(8)$$

In which

$$\Delta y_1 = hf(t_n, y_n)\#(9)$$

$$\Delta y_2 = hf\left(t_n + \frac{h}{2}, y_n + \frac{\Delta y_1}{2}\right)\#(10)$$

$$\Delta y_3 = hf\left(t_n + \frac{h}{2}, y_n + \frac{\Delta y_2}{2}\right)\#(11)$$

$$\Delta y_4 = hf(t_n + h, y_n + \Delta y_3)\#(12)$$

Here y_{n+1} is the RK4 approximation of $y(t_{n+1})$. y_{n+1} is obtained by the adding the weighted average of four increments to y_n. These increments are represented by equation (9) to (12). They are the product of the numerical interval h between t_{n+1} and t_n, and an estimated slope specified by function f on the right-hand side of the differential equation $f(t, y)$.

The specific steps of solution of the Bateman equation by the fourth-order Runge-Kutta method is:

$$I_{n+1} = I_n + \frac{1}{6}(\Delta I_1 + 2\Delta I_2 + 2\Delta I_3 + \Delta I_4)\#(13)$$

$$X_{n+1} = X_n + \frac{1}{6}(\Delta X_1 + 2\Delta X_2 + 2\Delta X_3 + \Delta X_4)\#(14)$$

In which

$$\Delta I_1 = \Delta t(\gamma_I \Sigma_f \phi - \lambda_I I_0)\#(15)$$

$$\Delta X_1 = \Delta t[\gamma_X \Sigma_f \phi + \lambda_I I_0 - (\lambda_X + \sigma_{aX}\phi)X_0]\#(16)$$

$$\Delta I_2 = \Delta t\left[\gamma_I \Sigma_f \phi - \lambda_I\left(I_0 + \frac{1}{2}\Delta I_1\right)\right]\#(17)$$

$$\Delta X_2 = \Delta t\left[\gamma_X \Sigma_f \phi + \lambda_I\left(I_0 + \frac{1}{2}\Delta I_1\right) - (\lambda_X + \sigma_{aX}\phi)\left(X_0 + \frac{1}{2}\Delta X_1\right)\right]\#(18)$$

$$\Delta I_3 = \Delta t\left[\gamma_I \Sigma_f \phi - \lambda_I\left(I_0 + \frac{1}{2}\Delta I_2\right)\right]\#(19)$$

$$\Delta X_3 = \Delta t\left[\gamma_X \Sigma_f \phi + \lambda_I\left(I_0 + \frac{1}{2}\Delta I_2\right) - (\lambda_X + \sigma_{aX}\phi)\left(X_0 + \frac{1}{2}\Delta X_2\right)\right]\#(20)$$

$$\Delta I_4 = \Delta t[\gamma_I \Sigma_f \phi - \lambda_I(I_0 + \Delta I_3)]\#(21)$$

$$\Delta X_4 = \Delta t[\gamma_X \Sigma_f \phi + \lambda_I(I_0 + \Delta I_3) - (\lambda_X + \sigma_{aX}\phi)(X_0 + \Delta X_3)]\#(22)$$

Where Δt is the time interval,
I_0 is the initial value of iodine, and
X_0 is the initial value of xenon.

The advantage of Runge-Kutta is that it applies the single-step method and has a high precision, and it is easy to change the step length when doing the calculation. The disadvantage is that it has to calculate the function value f for four times for each increment. Nevertheless, the amount of calculation is not a problem today due to the advanced computer technology. Therefore, the fourth-order Runge-Kutta method is one of the common methods that solve the differential equations.

The Fourth-order Runge-Kutta method is implemented in MATLAB to solve the Bateman equations for the iodine-xenon decay chain. Appendix A illustrates the algorithm employed in the MATLAB script to perform the calculation for xenon population transient of 70 time steps 3otalling 70 hours after reactor shutdown.

3.2 Matrix exponential method

In mathematics, the matrix exponential is a matrix function on square matrices analogous to the ordinary exponential function. It is used to solve systems of linear differential equations. MATLAB uses the Padé approximant to compute the matrix exponential [4]. It is applicable to solve the Bateman equations.

The general form of initial value problem of an ODE system is:

$$\frac{d\vec{x}}{dt} = A\vec{x}, \qquad \vec{x}(t_0) = \vec{x}_0\#(23)$$

In the equation, vector matrix $\vec{x}$ is a matrix that includes functions of time t, and the derivative of $\vec{x}$ equals to the product of $\vec{x}$ and a constant coefficient matrix A. The value of $\vec{x}$ is $\vec{x}_0$ at the initial time t_0.

The solution to the equation (23) is:

$$\vec{x}(t) = e^{(t-t_0)A}\vec{x}_0\#(24)$$

The vector matrix that need to be solved contains the functions of time with respect to the population of xenon and the population of iodine. Then, the constant coefficient matrix is calculated， after which the Bateman equation can be solved.

The solution of the initial value problem of the system of Bateman equation in equation (3) and (4) is:

$$\begin{bmatrix} I_{n+1} \\ X_{n+1} \end{bmatrix} = e^{\Delta t\begin{bmatrix} -\lambda_I I & 0 \\ \lambda_I I & -\lambda_X - \sigma_{aX}\phi \end{bmatrix}} \begin{bmatrix} I_0 \\ X_0 \end{bmatrix}\#(25)$$

The matrix exponential method as described by equation (25) is also implemented in MATLAB to solve the Bateman equation for the iodine-xenon decay chain to predict xenon population under reactor shutdown transient. The algorithm listed in Appendix A is also used to implement the matrix exponential to simulate xenon transient behaviour after reactor shutdown for 70 hours.

4 Benchmark study

Once implemented in MATLAB, the two methods, Runge-Kutta and the matrix exponential, are used to perform a benchmark test for a hypothetical reactor using realistic reactor parameter inputs. These parameters are tabulated in Table 1. The reactor is assumed to be homogenous and non-leaking, as assumed also in Section 2. The cross-section used are all one-group values.

Table 1. Reactor parameters: independent variables used in Bateman equation.

γ_I	Fission yield of I-135	0.061
γ_X	Fission yield of Xe-135	0.003
λ_I	Decay constant of I-135 (s^{-1})	2.874×10^{-5}
λ_X	Decay constant of Xe-135 (s^{-1})	2.027×10^{-5}
σ_{aX}	Microscopic absorption cross-section of Xe-135 (cm^2)	2.75×10^{-18}
ν	# of neutrons released per fission	2.3
ϕ	Thermal neutron flux ($cm^{-2}s^{-1}$)	4.42×10^{20}
Σ_f	Macroscopic absorption cross-section (cm^2)	0.008

In both methods, the function of poisoning in 70 hours after shutdown is investigated, so the total length is 252000 seconds, which is divided into 70 steps with a step length of 3600 seconds. In the hypothetical reactor being investigating, the thermal neutron flux is equal to 4.42×10^{20} neutrons per 4entimetre squared per second, and macroscopic absorption cross-section is equal to $0.008\ cm^2$.

By applying the two methods, the codes can investigate the xenon and iodine population. Then, the codes are able to calculate the poisoning. The function of poisoning used in the code is represented as:

$$\rho_{Xe} = -\frac{\sigma_{aX}X}{\nu\Sigma_f} \#(26)$$

The xenon poisoning level with respect to time after a reactor shutdown from different power levels simulated using the RK4 method is plotted in Figure 1. It is assumed that at t = 0 hour, the reactor (i.e. xenon population) is at equilibrium at the specified power level. The graph of xenon poisoning with respect to time is a curve that peaks very rapidly at first and then decays at a smaller slope. The graph show that the poisoning reaches its peak at around 11 hours after the shutdown and began to drop. It is also noticed that the higher the power level is before the reactor shutdown, the higher the peak of the xenon poisoning after shutdown.

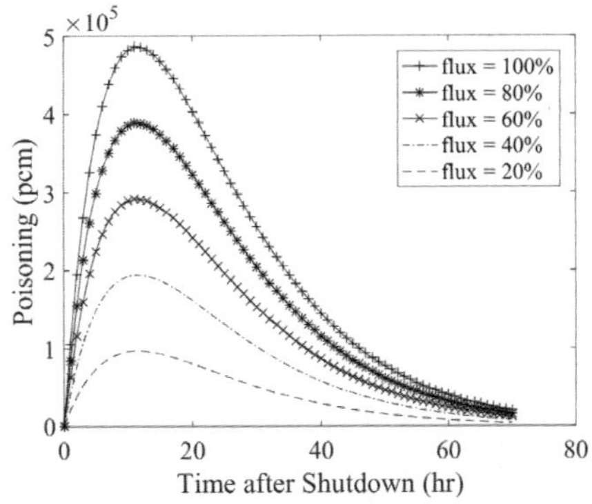

Fig. 1. Xenon Poisoning transient at different flux levels using the RK4 method.

The xenon poisoning transient result calculated using the matrix exponential method implemented in MATLAB. The two graphs show very similar pattern is plotted in Figure 2.

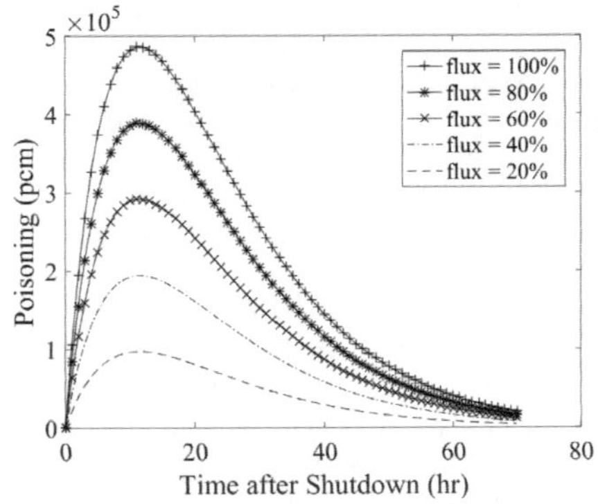

Fig. 2. Xenon Poisoning transient at different flux levels using the Matrix Exponential Method.

To compare the accuracy of the two solutions, the percentage error of each method with respect to analytical solution as presented in equation (6) is calculated. Figure 3 plots the percentage error of xenon poisoning of the full power case with respect to time after reactor shutdown using the RK4 method and Figure 4 plots the percentage error using the matrix exponential.

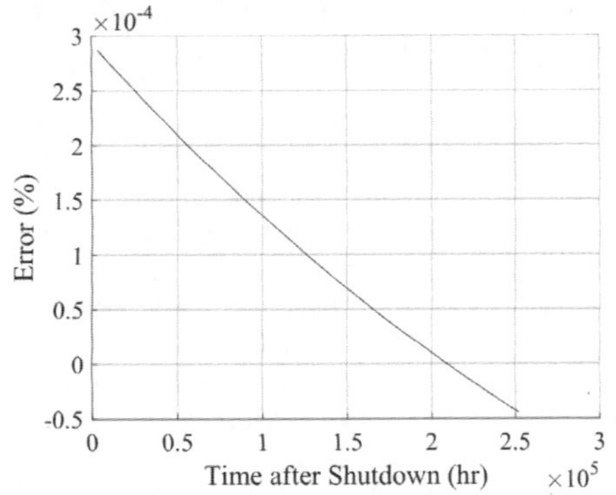

Fig. 3. Percentage Error of the RK4 Method.

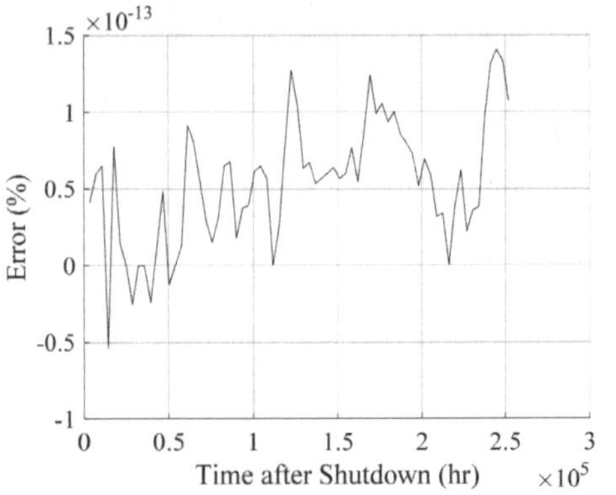

Fig. 4. Percentage Error of the Matrix Exponential Method.

As shown on the graphs above, the error of result from the RK method stays within 0.0003%. For the result generated by the matrix exponential method, the error stays within 1.5×10^{13}%. Clearly, the matrix exponential gives more accurate solutions.

The calculation time of each method is also tested to investigate the trade-off between accuracy and computational expense of the two methods. The average operation time of 300 runs is around 0.09 milliseconds using the RK4 method and is around 5 milliseconds when using the matrix exponential method.

Both methods are accurate and can provide good approximation of the "true" analytical answer. The matrix exponential method requires longer calculation time, but it is much more accurate than the Fourth-order Runge-Kutta method. The problem investigated at this stage of the project is relatively simple and both methods can provide accurate results within a very short time frame, therefore the matrix exponential is the preferred method. However, when the problem becomes more complicated, computational efficiency might become a much greater concern. The less computationally expensive method RK4 could become the appropriate.

5 Results and discussion

After the two models have been successfully benchmarked against the analytical solution, they are used to produce a xenon transient "surface" for different reactor operating conditions, i.e. neutron flux and one-group macroscopic fission cross-section. Figure 5 visualizes the 3D surface illustrating the shutdown transient of xenon poisoning reactor under different reactor equilibrium power levels before the shutdown. This essentially is Figure 1 presented in 3D space. In the calculation presented in Figure 1, a full power neutron flux of 4.42×10^{20} /s·m^2 as the value of neutron flux. A range of flux levels from 80% to 120% of the full power flux level with a 4% increment is explored. The importance of this "xenon transient surface" is that once the governing equation of the surface is known, the xenon population can be predicted using the surface equation without needing to re-calculate Bateman equation. This will speed up the design safety check process.

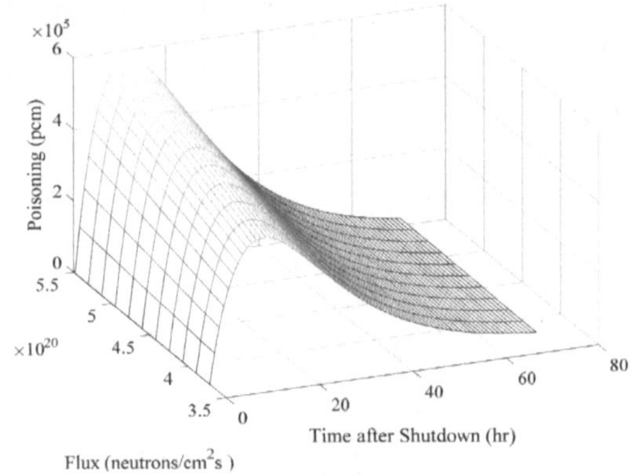

Fig. 5. Xenon Poisoning under Different Neutron Fluxes using the Matrix Exponential method.

Similar pattern can be seen in Figure 5 as seen in Figure 1 and 2. Xenon poisoning peaks around 11 hrs after reactor shutdown and then decays with a relatively slower rate. The height of the peak depends on the homogeneous one group neutron flux, in other words, the reactor power before reactor shutdown. The greater the flux is, the higher the poisoning peak is. The height of the xenon peak reaches almost 5×10^5 pcm at 100% power, meaning that a significant amount of reactivity is suppressed by the xenon poisoning. This means that there will be a time frame after which the reactor shutdown during which the reactor could not be restarted because of the lack of access reactivity. Even if the reactor restarts before the xenon reaches its peak, the continual production of xenon from iodine decay could still shut the reactor down. This is also one of the reasons why nuclear power plants normally operate as base load power provider, even during times of low electricity demand. Once the reactor shuts down, the core is theoretically impossible to restart until xenon poisoning decays to a level lower than the reactivity of the fuels, and the economic loss during the shutdown period can be significant.

Furthermore, power level fluctuations also lead to similar behaviours as discussed previously. When the core goes from a high flux state to a low flux state, xenon is also built up and supresses the core reactivity before reaching to a new equilibrium state at a lower poisoning level. During the xenon peak and decay, if the reactor were to be maintained critical, control rods would have to be gradually withdrawn to release more neutrons to the core during the xenon peak and then inserted again once xenon starts to decay. This makes reactivity control of the core extremely complicated and must be handled meticulously. The infamous Chernobyl accident was in fact caused by reactivity release of the decayed xenon poisoning [5].

 https://doi.org/10.1051/matecconf/201818601004

6 Appendix A

Algorithm used to implement the RK4 and the matrix exponential methods in MATLAB.

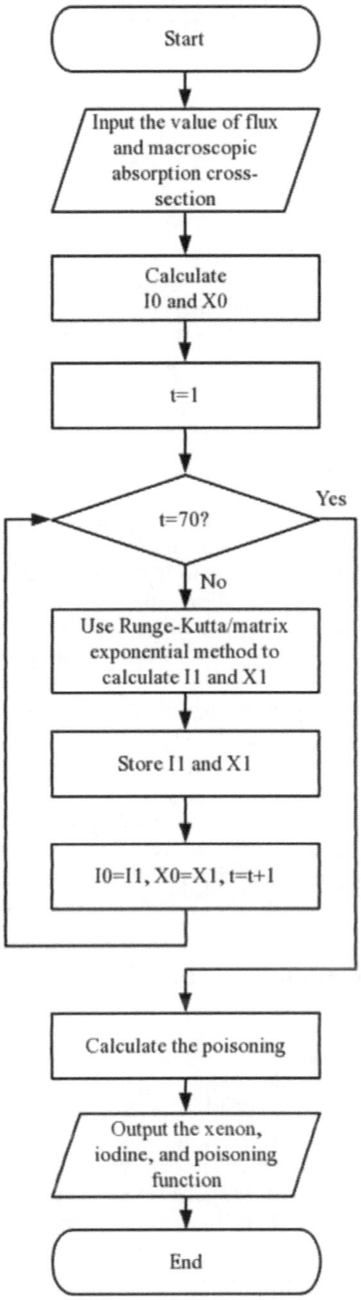

7 Conclusion and future work

Due to its extremely high absorption cross-section, xenon plays a vital role in reactor safety during operation; therefore, understanding xenon's transient behaviour under reactor power level change is of crucial significance for reactor design and operation. In this paper, two mathematical methods, i.e. Runge-Kutta and the matrix exponential, have been employed using MATLAB script to solve Bateman equations for the iodine-xenon decay chain to model xenon population behaviour under "reactor shutdown" transient scenarios. Results obtained from these two methods agree within an acceptable range with an error smaller than 3×10^{-4}%. The models are then used to test how the transient behaviour of xenon changes with respect to neutron flux, i.e. reactor power level. It is seen that with higher flux, a higher xenon poisoning peak is observed, longer time required to decay. The xenon build-up could render the reactor core impossible for restart during some time frame after a shutdown. Xenon fluctuation also complicates the reactivity control during operation if the core power level changes.

It must be acknowledged that the methods developed in this paper, however, can only provide rough estimations of xenon transient as they currently only solve one-group Bateman equations in homogeneous reactor environment. There are many factors that have not been included in the models that could affect the calculation results, such as spatial variation of nuclide distribution and cross-sections and energy level dependence of the parameters. Implementing these factors would require more sophisticated methods such as nodal diffusion methods, Method of Characteristics or Monte-Carlo neutron transport methods; however, as explained earlier, these methods require longer calculation time, especially when doing transient analysis and therefore might not be the best fit for preliminary reactor design stage. The simple models developed in this project can be modified in future work to include multi-group calculation, spatial heterogeneity and couple to legacy codes to perform xenon transient analysis on the fly of reactor design with more detailed reactor specific information.

References

1. E.E. Lewis, *Fundamentals of Nuclear Reactor Physics*, Chapter 10, pp. 243-258, Academic Press (2008).
2. DEVRIES, Paul L., HASBUN, Javier E., *A first course in computational physics*, Second edition, pp. 215, Jones and Bartlett Publishers (2011).
3. Joe D. Hoffman, *Numerical Methods for Engineers and Scientists*, pp. 370-372, Marcel Dekker (2001).
4. Higham, N. J., *The Scaling and Squaring Method for the Matrix Exponential Revisited*, pp. 1179–1193, SIAM J. Matrix Anal. Appl., **26(4)**, (2005).
5. C.D. Bowman, *What Happened at Chernobyl?*, Science (New York, 2011).

Natural Qubit Matrix of Primary Elements of Matter

Eugene Machusky

[1]National Technical University of Ukraine "Kyiv Polytechnic Institute", Kyiv, Ukraine

Abstract. The arithmetic limits of natural bit calculations are strictly established. The natural quantum metric system has been developed. Only seven scaling units that generate thirteen invariant values are necessary and sufficient for an accurate estimation and a mutual coordination of the fundamental constants of quantum physics. For the first time, the calibration constants of quantum physics were obtained, calculated and identified analytically with almost an absolute accuracy, which is limited only by a bit capacity of a computer. The basic constants of quantum physics are, in fact, the dynamic parameters of the functional relationships of the transcendental numbers PI and E with natural numbers, which draw a holographic picture of the motion of spherical wave fronts.

1 Introduction

The first attempt to develop the absolute metric system was made by Gauss in the early nineteenth century. An attempt was unsuccessful due to the inconsistency of parabolic, elliptical and hyperbolic geometry. The inverse logarithmic relationships of the transcendental numbers PI and E resolve this problem and create the absolute metric without any artifacts and without measurements at all. It suffices to postulate the number PI as the universal unit of space and the number E as the universal unit of time. Then the first derivative of PI by E can be considered as a unit of speed, and the second derivative is the unit of acceleration. Geometrically, PI is the perimeter; E is the diameter of the space-time sphere. E/2 is the radius, 4*PI*(E/2)^2 is the surface of the sphere, and 4/3*PI*(E/2)^3 is the volume of the sphere, PI*(E/2)^2 is the surface of a circle, N*(PI*E^2)/4 is a discrete set of the cylinder volumes. A functional relationships between these space-time geometric parameters and an infinite set of integers N constitute the basis of the natural [E...PI] binary calculations of the parameters of wave motion [1-5].

The basic principles of universal arithmetic and geometry were developed by Newton, Leibniz and Euler at the beginning of the eighteenth century. The principles led to the creation of the differential calculus and analysis: Unit 1 generates a natural set: 1+1+1+... = 1... (N+1). There is no absolute infinity. Unit 1 and a natural set generate the inverse natural set: 1/(1+1)...1/(N+1). There is no absolute zero. Unit 1 is a quantum of the progressive natural set. Units 1/(N+1) are the quanta of the inverse natural set. An inverse natural set generates a universal spatial unit PI = (1+1)*Sum{d(1/N)/Sqrt(1-1/N^2)} as a circle perimeter. An inverse natural set generates a universal temporal unit E = (1+1/N)^N as the base of natural logarithms.

The modern science has emerged after the discovery of logarithms. But it was impossible to resolve the fundamental problems of infinitesimal calculus before the discovery of the laws of thermodynamics and electrodynamics, from which an approximate scaling unit of temperature K (kelvin) and a scaling unit of the rotational speed C (the speed of light) were obtained. It will be shown below that K and C are essentially fundamental mathematical units of the inverse natural calculus, such as PI and E of the standard analysis, and they can be obtained analytically from the mutual functional relations of the transcendental numbers PI and E. As the result, all other constants of quantum physics can also be obtained analytically, and their physical meanings can be explained geometrically.

2 Arithmetical limits of natural computations

Any two numbers can be a basis of the calculation system, for example binary [0...1], decimal [0...9], quantum binary [0.000>...1.111>] or natural binary [E...PI]. The functional relations of two transcendental numbers PI and E with three natural numbers A = 137 (Gaussian prime), B = 602214183 (Avogadro's integer) and R = 105456978 (integer of Dirac) completely describe a harmonic motion of spherical waves and create the system of natural computations in relative space-time coordinates.

A square of the normalized nine-digit qubit [1.111111111]^2 = 1.2345678987654321 has a 17-bit mirror symmetry.

A square of the normalized infinite qubit [1.11111111 ...]^2 = 1.23456790123456790 has a 9-bit progressive symmetry.

A cube of the normalized infinite qubit connects fields of irrational and rational numbers by truncated 17-bit sequence:

[1.11111111>]^3 = 1.3717421124828532235939643470...,
[1.11111111>]^3 = Sum {[137+(137-100)*N]/10^(3*N+2)} - natural notation,
[1.11111111>]^3 = [1+1/10+1/100+1/1000+...]^3 - binary-decimal notation,
[1.11111111>]^3 = [10^0+10^(-1)+10^(-2)+10^(-3)+...]^3 - decimal notation,
[1.11111111>]^3 = [10/(10-10^0)]^(10^0+10^0+10^0) - zero-decimal notation,
[1.11111111>]^3 = [(2*5)/((2*5)-1)]^3 - first prime notation,
[1.11111111>]^3 = (2^3)*(5^3)/(3^(2*3)) - second prime notation,
[1.11111111>]^3 = (2*2*2*5*5*5)/(3*3*3*3*3*3) - third prime notation,
[1.11111111>]^3 = (1000)/(729) = 1.3717421124828532.

The relation 1/A = (10^0)/137 = Sum {[729927]/10^(8*N)} is the first hyperbolic point of the periodic mirror supersymmetry of the inverse natural set. For any decimal digit a, b, c, d, the ratio [abcdabcd] / [abcd] = 10001 = 137*73 = 137*(137-64).

At the same time, the number A = 137 is very close to the squared sum of root mean the square, arithmetical, geometrical, harmonic mean of PI and E:

[Sqrt((PI^2+E^2)/2)+(PI+E)/2+Sqrt(PI*E)+2*PI*E/(PI+E)]^2 = 136.9938985020083597 = SMS.

SMS sequence can be treated as the entropy bound of the parabolic coupling of binary, decimal and natural bit computation systems. The approximate equations E^137= 100*PI*10^57 and E^137 = 100*Sqrt(10)*10^57 can be treated as the exponential entropy bounds of polynomial coupling of binary, decimal and natural calculation systems:

100*PI*10^57 = E^136.9972503724980956,
A^1 = 137.0000000000000000,
100*Sqrt(10)*10^57 = E^137.0038130331457184.

The computational links of the finite natural difference [PI-E], of the finite natural differential ND = [PI-E]*[PI/E]*[E/PI] and of the finite natural median NM = Sqrt(SMS)/4 are shown below:

TRUNCATED PI	TRUNCATED E	64-BIT ND	17-BIT NM
3.1415926535897932	2.7182818284590452	0.42331 082513 07479...999	2.9261098161852234
3.141592653589793	2.718281828459045	0.42331 082513 07479...999	2.9261098161852232
3.14159265358979	2.71828182845904	0.42331 082513 07499...999	2.9261098161852192
3.1415926535897	2.7182818284590	0.42331 082513 06999...999	2.9261098161851550
3.141592653589	2.718281828459	0.42331 082512 99999...999	2.9261098161848172
3.14159265358	2.71828182845	0.42331 082512 99999...999	2.9261098161758055
3.1415926535	2.7182818284	0.42331 082509 99999...998	2.9261098161112635
3.141592653	2.718281828	0.42331 082499 99999...999	2.9261098156624848
3.14159265	2.71828182	0.42331 082999 99999...999	2.9261098100647471
3.1415926	2.7182818	0.42331 079999 99999...999	2.9261097755621123
3.141592	2.718281	0.42331 099999 99999...999	2.9261090710239372
3.14159	2.71828	0.42331 000000 00000...000	2.9261075871661980
3.1415	2.7182	0.42329 999999 99999...999	2.9260226569083487
3.141	2.718	0.42299 999999 99999...999	2.9256776292410519
3.14	2.71	0.42999 999999 99999...999	2.9210438087537571
3.1	2.7	0.40000 000000 00000...000	2.8965476238998343
3	2	1.00000 000000 00000...000	2.4747498748948926

The 6-bit integer 42331 and the 16-bit integer 4233108251307479 are prime numbers. This fact limits an accuracy of computations of the hyperbolic functions Cosh and Sinh in the decimal calculation system with floating point.

The characteristic feature of natural differentials (ND) is an appearance of very long sets of digits 9. It is obvious that 0.9999 ... = 1 and 9.999 ... = 10. An infinite sequence of 9 is generated as the integer B = 602214183 and the integral of Avogadro as BS = Sum {B/10^(3*N+11)} = 0.0060281699999999 = 0.00602817.

The characteristic feature of natural medians (NM) is an appearance of the "golden sequence" 1618 over 12-bit [PI...E]. This fact limits an accuracy of the natural median computations of harmonic functions Cos and Sin in decimal system with floating point.

It is easy to show that the use of symbolic infinite (pi...e) and truncated decimal [PI...E] notations leads to essentially different arithmetic results of calculations of polynomial and exponential functions:

1/pi^32 = 1/e^37 = 1/10^16,
1/pi^33 = 1/e^38 = 0.
1/PI^32 = 1234/10^20 > 1/E^37 = 853/10^20,
1/PI^33 = 393/10^20 > 1/E^38 = 314/10^20.
1/PI^129 = 1/E^148 = 1/NM^137 = 1/10^64,
1/PI^130 = 1/E^149 = 1/NM^138 = 0.

e^(j*pi) + 1 = 0 in symbolic (pi...e) continuous mathematics,

E^(j*PI) + 1 = 1/10^64 > 0 in binary [0...1] discrete mathematics,

E^(j*PI) + 1 is about [1/137]/10^57 > 0 in qubit [0>...1>] arithmetic.

3 Natural quantum metric

Seven scaling units are sufficient to constitute the natural metric system. The numbers PI and E are the basis of natural differential computing, the numbers C and K are the basis of a natural median computing. The integers R, A, B mutually co-ordinate the natural differentials (ND) and the natural medians (NM):

PI=(1+1)*Sum{d(1/N)/Sqrt(1-1/N^2)}= 3.1415926535897932 (spatial unit of Pythagoras),

E=(1+1/N^N)^(N^N)=2.7182818284590452 (temporal unit of Euler),

C=(R/10^8+4*PI*C/10^18)^64*10^7= 299792457.86759134 (speed unit of Maxwell),

K=E+AS+BS=2.7315999984590452 (temperature unit of Kelvin),

R/10^8=Integer{10^8*(C/10^7)^(1/64)}/10^8= 1.0545697800000000 (radial unit of Dirac),

AS=1/100/Sum{[A+(A-100)*N]/10^(3*N+2)}= 0.0072900000000000 (eccentricity unit of Sommerfeld),

BS=Sum{B/10^(3*N+11)}=0.0060281699999999 (entropy unit of Avogadro).

4 Matrix equations of wave motion

Ten matrix equations of the differential geometry in a topologically inverted space-time determine parameters of the push-pull spherical wave motion:

[R]=1+2/100*(E+[A]*(1+Sqrt(2*PI*E/100))) – matrix field of the normalized inverse radius (Dirac's units).

[A]=(100*([R]-1)/2-E)/(1+Sqrt(2*PI*E/100)) – matrix field of the normalized inverse eccentricity (Sommerfeld's units).

[P]=2*PI*[R] – matrix field of the normalized inverse perimeter (Planck's units).

[G]=[P]*(1+[A]) – matrix field of the normalised density of inverse perimeters (Newton's gravity units).

[V]=[R]^64*10^7 - matrix field of the relative speed of waves (Maxwell's units).

[MM]=12-[A]/10 - matrix field of the relative amplitude of waves (Amagat's molar concentration units).

[KB]=Cos[MM]-Sin[MM] – matrix field of the relative phase of waves (Boltzmann's units).

[W]=[V]^2 – matrix field of the relative energy of waves (Einstein's units).

[NA]={Sqrt(8*PI*E/(8*PI*E+137^2))/(1+2*[A]/1000)-5/10^8}/10 – field of Avogadro entropy units.

[DA]=1/[NA]/100 – normalized inverse information entropy (Dalton's atomic mass units).

5 Matrix of quantum invariants

The arithmetical entropy matrix of Avogadro-Dalton:
A4=(PI*E/100)^2+(1/137-(PI*E/100)^2 is upper parabolic bound of the inverse orbital eccentricity.
AH=1/(16*PI*E) is upper hyperbolic bound of the inverse orbital eccentricity.
AB=A(NB), NB = 602214183/(1+4*PI/10^8)/10^11 is PI-reference point of the information entropy of eccentricity.
AL=1/(1+59*Ln(10)) is upper logarithmic bound of inverse orbital eccentricity.

[N]	[A]	[R]	[DA]
0.0060221410025819	0.0073189621138002	1.0545725104198716	1.6605390002845580
0.0060221410053903	0.0073187289405399	1.0545725038291215	1.6605389995101736
0.0060221410732354	0.0073131309589000	1.0545723455995802	1.6605389808026354
0.0060221411450152	0.0073071361524362	1.0545721761536017	1.6605389610101475

The geometrical entropy matrix of Planck-Sommerfeld:

A1=1/137 is central point of natural mirror symmetry of inverse orbital eccentricity. AF = 1000/Integer {1000* Sqrt (137^2+PI^2)} is PI-reference point of inverse orbital eccentricity.

A0 = (PI*E/100)^2 is lower parabolic bound of orbital eccentricity.

AS = 1/100/(1.111111111>)^3 is lower hyperbolic bound of orbital eccentricity.

[N]	[A]	[R]	[P]
0.0060221412397551	0.0072992700729927	1.0545719538152265	6.6260710055755005
0.0060221412628504	0.0072973525205056	1.0545718996147182	6.6260706650236630
0.0060221413188129	0.0072927060593902	1.0545717682803448	6.6260698398254579
0.0060221413514050	0.0072900000000000	1.0545716917923240	6.6260693592370495

The differential entropy matrix of Kelvin-Maxwell

RC = (R+4*PI*C/10^10)/10^8 is the PI-reference point of core radius. RE = (R+1/E)/10^8 is the E-reference point of core radius.

RA = (R+1/(E+AS))/10^8 is the A-E reference point of core radius.

RK = (R+1/(E+AS+BS)/10^8 is the K-reference point of core radius.

[KB]	[A]	[R]	[V]
1.3806484502840000	0.0072224962396475	1.0545697837673031	299792457.86759134
1.3806484501880000	0.0072224931083101	1.0545697836787944	299792456.08618159
1.3806484501770000	0.0072224927601997	1.0545697836689549	299792456.07825451

1.3806484501680000 0.0072224924737441 1.0545697836608581 299792455.93094320

The median point of a normal distribution of the Wien wavelength displacement:

AX=5/Root{X*E^X/(E^X-1)=5}-1 is the lower bound of the relative inverse eccentricity:

[NX]	[AX]	[RX]	[GX]
0.0060221445289294	0.0070261763632109	1.0545642346951568	6.6725781076198224

6 Primary physical elements of matter

Background tempo of the core waves:
E+AS+BS = 2.7315999984590452 = K is the upper bound of the background tempo.
Med {EA...K} = 2.7285850810946404.
E+AS = 2.7255718284590452 = EA.
Med of meds = 2.7252543275634558 = TBG is the median background tempo.
Med {EA...E} = 2.7219256081809303.
E+0 = 2.7182818284590452 = E is the lower bound of the background tempo.

Vibrational speed of the core waves:
RE^64*10^7 = 2.9979245625727419*10^8 = TE.
Med {TA...TE} = 2.9979245616776434*10^8.
Med of meds = 2.9979245608618159*10^8 = TN+.
Med {TN-...TN+} = 2.9979245608610649*10^8 = T0 is the median vibrational speed.
R+1/TBG/10^8)^64*10^7 = 2.9979245608603140*10^8 = TN-.
RA^64*10^7 = 2.9979245607825451*10^8 = TA.
Med {TA...TK} = 2.9979245600459885*10^8.
RK^64*10^7 = 2.9979245593094320*10^8 = TK.

Rotational speed of the core waves:
CN+ = R/10^8+4*PI*C/10^18)^64/10 = 2.9979245786759134*10^8.
Med {CN-...CN+} = 2.9979245786759103*10^8 = C0 is the median rotational speed.
CN- = RC^64/10 = 2.9979245786759074*10^8.

Translational velocity of the core waves:
Med {T0...C0} = 2.9979245697684876*10^8 = V0 is the median translational velocity.

Median radius of the core waves:
(V0/10^7)^(1/64) = 1.0545697837183448 = RV0.

Median eccentricity of the core waves:
(100*(RV0-1)/2-e)/(1+Sqrt(2*PI*E/100)) = 0.0072224945075574 = AV0.

Relative amplitude of the core waves:
12-AV0/10 = 11.9992777505492443 = MM (molar mass)

Relative phase of the core waves:
Cos(MM)-Sin(MM) = 1.3806484502310000 = KB (Boltzmann's unit).

Density of the nuclear waves:
2*PS*(1+AS) = 6.6743734048658876 = GS
Med {GV0...GS} = 6.6741437171665912
Med of meds = 6.6736948779387962 = GM is the lower bound of the gravitational unit.
GV0 = 2*PI*RV0*(1+AV0) = 6.6739140334195273
Med {GV0...GX} = 6.6732460538038305
2*PX*(1+AX) = 6.6725781076198224 = GX

Parameters of the orbital waves:
R1 = 1.0545719538152265, 2*PI*R1 = 6.6260710055755005 = P1.
Med {P0...P1} = 6.6260704227004664.
Med of meds = 6.6260700111158524 = PP is the median perimeter (Planck's unit).
R0 = 1.0545717682803448, 2*PI*R0 = 6.6260698398254579 = P0.
Med {P0...PS} = 6.6260695995312515.
RS = 1.0545716917923240, 2*PI*RS = 6.6260693592370495 = PS.
PP/(2*PI) = 1.0545717955420578 = RP is the reduced Planck unit.
(100*(RP-1)/2-E)/(1+Sqrt(2*PI*E/100)) = 0.0072936705478647 = AP.
1000/Integer{1000*Sqrt(137^2+PI^2)} = 0.0072973525205056 = AF is the fine structure constant.
1+2/100*(E+AF*(1+Sqrt(2*PI*E/100))) = 1.0545718996147182 = RF.

The median density of the orbital waves
Med {RP...RF} = 1.0545718475783874 = RQ.
(100*(RQ-1)/2-E)/(1+Sqrt(2*PI*E/100)) = 0.0072955115341644 = AQ.
2*PI*RQ = 6.6260703380697541 = PQ.
PQ*(1+AQ) = 6.6744109106473266 = GQ is the upper bound of the gravitation unit.

The median orbital-nuclear density of waves
Med (GM...GQ) = 6.6740528894917918 = G is a median of the Newtonian gravitational unit.

The elementary charge unit:
Sqrt(10*AF*RP/V0) = 1.6021766150249783 = Q is the Stoney elementary charge unit.

The median information entropy:
N(A4) = 0.0060221410025819 = N4.
Med {NH...N4} = 0.0060221410039861 = NH4.
N(AH) = 0.0060221410053903 = NH.
Med {NHL...NH4} = 0.0060221410395944 = NE.
Med {NB...NE} = 0.0060221410564149 = NA is the median of the Avogadro entropy unit.
NB = 0.0060221410732354.
Med {NH...NL} = 0.0060221410752027 = NHL.
N(AL) = 0.0060221411450152 = NL

Decimal orders of quantum units are easily extracted from the gauge equation E^137 = PI*10^59 and from the Wien wavelength displacement law WD = 2*PI*RP*(T0...C0)/(KB*X). A natural computing fully coordinates classical and quantum physics:

Mechanical energy quanta	EM = (m*v^2/2)*[1+(v^2)/(T0...C0)^2], (from Newton to Einstein).
Thermal energy quanta	ET = (k*T)/[1+(v^2)/(T0...C0)^2], (from Kelvin to Boltzmann).
Electrical energy quanta	EE = (q*U)/[1+(v^2)/(T0...C0)^2], (from Stoney to Dirac).
Optical energy quanta	EA = (h*f)*[1+(v^2)/(T0...C0)^2], (from Planck to Shannon).

References

1. Eugene Machusky, Quantum Metric of Classic Physics, (2017), iopscience.iop.org/article/10.1088/1757-899X/239/1/012009.
2. Eugene Machusky, Unified Quantum Metric, (2017), J Laser Opt Photonics 4:3 (Suppl) DOI:10.4172/2469-410X-C1-014.
3. Eugene Machusky, Quantum Physics as Dynamic Space-Time Differential Calculus, (2017), J Material Sci Eng 6:7 (Suppl) DOI:10.4172/2
4. Eugene Machusky, Quantum Physics as Dynamic Space-Time Differential Calculus, review article (2017), Adv. Material Sci Engg, volume 1, issue 1, 1 of 4. http://www.opastonline.com/wp-content/uploads/2017/12/quantum-physics-as-dynamic-space-time-differential-calculus-amse-17.pdf
5. Eugene Machusky, Complex Geometry of Wave Motion, (2018), Int J Eng Tech DOI: 10.7763/IJET.2018.v10.1056.

An Overlapping and Integral Blurry Evaluation Method to Optimize Tissues Scanning from Microscope

Jaishir Bayuelo[1], *Javier* Sanjuan[2], *Julián* Yepes-Martinez[2], *Wilson* Tovar[1], *Fabio* Zapata[1], *Giselle* Peñaloza[2]

[1]Department of Systems Engineering, Universidad del Norte, Barranquilla, Colombia

[2]Department of Mechanical Engineering, Universidad del Norte, Barranquilla, Colombia

ABSTRACT. Virtual microscopes are devices that employ an automated XYZ mechanism to scan a sample, leading to the obtention of a series of small pictures that, when merged, compose a high-quality representation of the specimen. Due to the assembly tolerances, these devices may suffer from zones out of focus, reducing the quality of the final image. To solve this problem, researchers employ evaluation methods to calculate the blurriness of the image, and when an out of focus picture is located, performs the process of autofocus. Because of the variation on the types of samples, especially in pathology, the existing evaluation methods may fail to deliver a proper blur detection. This article proposes an optimized algorithm for the detection of the blurriness while conducting the sample scan in real time, ensuring that every scanned picture will be in focus. For this purpose, the algorithm relies on two functions, the comparison of the overlapping zones of two consecutive images, and the multivariate linear regression of a series of focus functions. The algorithm proved to be a reliable tool when applied in different pathology samples.

1 Introduction

Whole-slide-Image (WSI) is a procedure that permits the digitalization of samples of pathological purpose to obtain high-resolution images. Due to the various applications of this technology, it has become a key factor for the innovation of digital pathology [1], unleashing in the construction of different microscopes based on this approach [2-11]. Unfortunately, there are limitations for the implementations of WSI for clinical use, among them is its high cost due to the use of specialized software, hardware, and maintenance .

Various microscopes have been proposed to tackle the mentioned drawback. Notwithstanding, the proposed low budget devices decrease the image quality of the sample. Due to the tilt of the mechanical stage produced by the increasing assemble tolerances, that causes blurriness in small zones of the WSI. This inconvenient, although on a lesser scale, is still an issue of expensive microscopes [12]. Different researchers have stated algorithms to detect the blurry zones and to conduct the autofocus of the microscope. The detection of the blurry zones is accomplished by the selection of an adequate focus measure. Many subroutines were found in the literature to calculate the focus, such as Tenengrad function, energy gradient function, Brenner function, and Entropy function [13-21]. The evaluation method should be selected based on its performance, unbiasedness, higher signal to noise ratio [11][18][22]. Unfortunately, the wide variety of tissues make the selection between those functions difficult, as for some samples the functions can detect false blurriness.

To increase the generality of the evaluation function we propose a new algorithm, overlapping and integral blurry evaluation method (BIEMO). This method integrates the contrast measure function, variance function, the lens distance, the Laplacian variance, and the Brenner function, to calculate the focus of the microscope. Furthermore, as the WSI is based on the capture of a series of small pictures and then in the stitching of them, we also use this stitching process as a second evaluation function. The stitching process merges two subsequent images calculating the feature difference between the overlapping regions of the two pictures. Thus, we propose the use of this overlapping region as a mean to evaluate the degree of blurriness of two images.

2 Functions and correlations

Let's consider the WSI image, which is divided by a set of small pictures C(i), each figure being in gray-scale. Then, for each pair of adjacent photos, a series factors are considered to determine if an image is out focus. The mentioned factors are the key-points matcher Km, the data variance σ, the Brenner's High Bhq and low Blq quality scores, the contrast measure function Cm, the Laplacian variance of the matrix L_V, and the distance between the camera and the sample F_d. The definition of these functions is in the following section.

2.1 Functions definition

The key-points matcher Km is the function utilized for the overlapping. Given two images, the function searches for the interest points to locate the area that is coincident in both pictures, returning a number that quantifies the percentage of coincidence. The algorithm used in this article for the calculation of the key-points is the AKAZE key-point detector and the Flann Based Matcher. Due to the complexity of those functions, the reader is encouraged to look for the reference [23].

The mean and the variance functions are both expressions used to evaluate the blurriness of the image (in the next sections we will be using N and M as the dimensions of the matrix, where N is number of columns and M number of rows). They are presented in the following:

$$u(I) = \frac{1}{NM}\sum_{i=1}^{N}\sum_{j=1}^{M} I(i,j) \quad (1)$$

$$\sigma(I) = \frac{1}{NM}\sum_{i=1}^{N}\sum_{j=1}^{M}[I(i,j) - u(I)]^2 \quad (2)$$

The Brenner's high (B_{hq}) and low (B_{lq}) quality scores calculate how much of an image has high and low quality, we calculate the proportion of the two quantities B_{hq}/B_{lq}. The following equation defines both quality scores:

$$B_{hq}(I) = \sum_{i=1}^{N}\sum_{j=1}^{M-4}\frac{1}{4}\left[\sum_{k=j}^{j+1} I(i,k) - \sum_{k=j+2}^{j+3} I(i,k)\right]^2 \quad (3)$$

$$B_{lq}(I) = \sum_{i=1}^{N}\sum_{j=1}^{M-6}\frac{1}{4}\left[\sum_{k=j}^{j+1} I(i,k) - \sum_{k=j+3}^{j+5} I(i,k)\right]^2 \quad (4)$$

The main feature of this Brenner coefficient is that when the image is focused, the value is lesser than 1 and when is not focused is greater or equal to 1. Figure 1 has a Brenner coefficient of 0,59.

- The Laplacian variance (equation 5) highlights the edges of the image, using the gradients in both directions of the Cartesian plane. We rely on this measure to calculate how blurry is the photo. Then, if we cannot differentiate the edges of the image, the picture is out of focus. Figure 2, which has a Laplacian variance of 118,68 and a normal variance of 1004,65. Thus, due to the difference between the two measures is huge, we conclude that the picture is out of focus.

$$LP(I) = \frac{\partial^2 I}{\partial x^2} + \frac{\partial^2}{\partial y^2} \quad (5)$$

- To calculate the contrast measure Cm, we calculated The Sobel operators. The Sobel operators are a measure of the robustness of the image to noise. To calculate them we calculate two gradients on the image, Gx and Gy, by applying the following kernels, respectively.

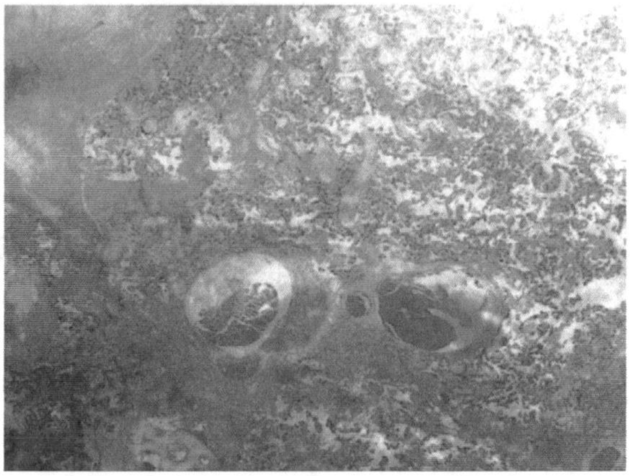

Fig. 1. Kidney tissue image token from 68,001 mm of the lens.

Fig. 2. Kidney tissue image token from 73,049 mm of the lens.

$$Kernels\colon x - = \begin{bmatrix} -1 & 0 & 1 \\ -2 & 0 & 2 \\ -1 & 0 & 1 \end{bmatrix}, y - = \begin{bmatrix} -1 & -2 & -1 \\ 0 & 0 & 0 \\ 1 & 2 & 1 \end{bmatrix} \quad (6)$$

Then, by applying the norm to the gradients, we obtain the contrast measure utilizing the next equation:

$$C_m = \frac{\sqrt{|G_x{}^2| + |G_y{}^2|}}{NM} + 10^{-6} \quad (7)$$

2.2 Multivariate linear regression

We utilize the functions defined in section 2 to obtain a multivariate linear regression model to predict the focus of the picture. We picked those functions due to the results of the ANOVA analysis, which stated that these evaluation methods had the greatest correlation with the blurriness of the image. The equation of the model is the following:

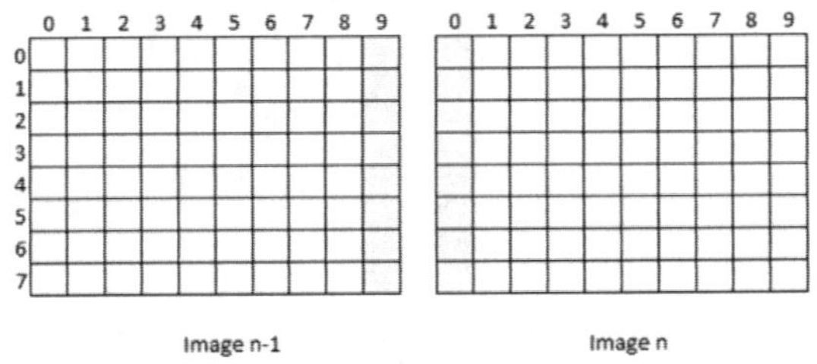

Fig. 1. Image matrices comparison

$$y = \frac{2{,}81C_m}{10^4} - \frac{5{,}46\sigma}{10^5} - \frac{1{,}09F_d + 2{,}49L_v}{10^3} + 1{,}8\frac{B_{hq}}{B_{lq}} + 2{,}14 \quad (8)$$

Where a value of y close to 1 means that the image is focused, otherwise it means blur. We obtained the previous equation employing a dataset of 100 images for the linear regression. Moreover, to prove the effectiveness of the model, we calculated the determination coefficient R, which yielded a value of 0, 98, implying fitness of the equation to describe the blurriness.

3 Biemo algorithm

Figure 4 presents pseudo-code of the BIEMO algorithm. The algorithm starts at the first position, C_{i-1}, then goes to the next point, C_i (with i starting in 1). With both of the images in gray-scale, the algorithm divides each image into ten equally spaced parts. Then, the algorithm compares the last division area (LDA) of the C_{i-1} image with the first division area (FDA) of the image C_i. Figure 3 describes the process. The FDA and LDA are used to calculate Km if this value is below to 0.001 the microscope must conduct autofocus; otherwise, the microscope goes to the next image C_{i+1}. If the value of i is less than the total of images n, the algorithm calculates i mod 6. If this operation is equal to 0 the algorithm employs equation 8, then if this value is less or equal to 0.5 the microscope conduct autofocus. In the opposite case, the algorithm returns to step 2. The i mod 6 function was implemented to reduce the use of the regression (decreasing the computational cost) and to avoid the mix between different zones of tissue, in the case of pathological samples.

4 Experiment and analysis

We captured six sequential images to test the validity and feasibility of the algorithm. To analyze the different scenarios in the analysis some of the pictures were out of focus, and others in the transition zones of different tissues.

For the analysis we use two matrices, a state matrix (Tables 1,3) and a comparison matrix (Tables 2,4). In the analysis, we replace a focused image by a blurred one in order to see how much the key-points matches decreased (we did it on purpose). Two different results were found (Tables 1-4).

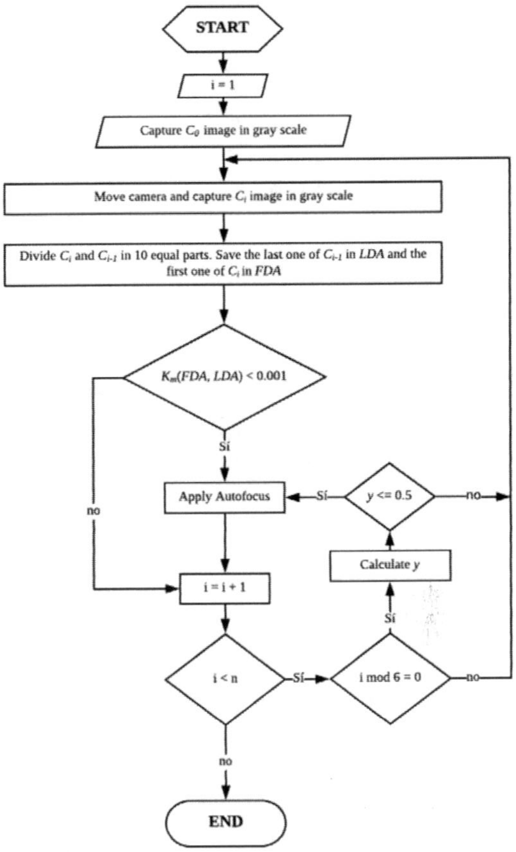

Fig. 2. Algorithm flow chart

- **State matrix:** This matrix consists of 6 columns, image number, focused, first division, last division, first division area and last division area.

(i) Image number: unique id for the image in process.
(ii) focused: YES/NO field, describe if the image is focused or not.
(iii) first division: describe if the first division of the image is dense or has dense tissue in the area, i.e compact.
(iv) last division: describe if the last division of the image is dense or has dense tissue in the area, i.e compact.
(v) first division area: total area of the first division image in square pixels.
(vi) last division area: total area of the last division image in square pixels.

- **Comparison matrix**

(i) image 1: id of the first image of the pair to compare key-point matches.

Table 1. State matrix 1

State Matrix 1					
Image	Focused	First Division	Last Division	FDA	LDA
0	YES	DENSE	DENSE	128271	128271
1	NO	DENSE	DENSE	128271	128271
2	YES	NO DENSE	DENSE	128271	128271
3	YES	DENSE	DENSE	128271	128271
4	YES	DENSE	DENSE	128271	128271
5	YES	DENSE	DENSE	128271	128271

Table 2. Comparison matrix 1

Comparison Matrix 1					
Img 1	Division	Img 2	Division	Km	Proportion
0	First	1	Last	9	3,5082E-05
1	Last	2	First	4	1,5592E-05
2	Last	3	First	821	0,003200256
3	Last	4	First	1463	0,00570277
4	Last	5	First	891	0,003473116

Table 3. State matrix 2

State Matrix 2					
Image	Focused	First Division	Last Division	FDA	LDA
0	YES	DENSE	DENSE	128271	128271
1	YES	DENSE	DENSE	128271	128271
2	YES	NO DENSE	DENSE	128271	128271
3	YES	DENSE	DENSE	128271	128271
4	YES	DENSE	DENSE	128271	128271
5	YES	DENSE	DENSE	128271	128271

(ii) *division*: division of the image in process (first, second, .. ,tenth or last).
(iii) *image 2*: id of the second image of the pair to compare key-point matches.
(iv) *Km*: number of key-point matches between last division of image 1 and first division of image 2.
(v) *proportion*: the relation between the number of key-point matches and the sum of the areas of the last division area of image 1 and the first division area of image 2.

From the results obtained in Tables 1-4, we conclude that any change in the blur level of the study area will make the algorithm to do the auto-focus process. In Table 2, we can see a proportion colored in red, as well as in Table 4, that means that in Comparison matrix 1, the proportion of key-point matches decreased by 94% with respect to Comparison matrix 2, due to the blur image that we introduce in the first experiment (replacing the focus version by the blur version). In this way, if we continue analyzing in group of six images, we will get a high performance and less computational cost of processing, because we will use the regression model each six times instead of 1 and detect in real time when the camera get out of focus based on this two factors.

Table 4. Comparison matrix 2

Comparison Matrix 2					
Img 1	Division	Img 2	Division	Km	Proportion
0	First	1	Last	152	0,000592496
1	Last	2	First	4	1,5592E-05
2	Last	3	First	821	0,003200256
3	Last	4	First	1463	0,00570277
4	Last	5	First	891	0,003473116

5 Conclusions

In this article was introduced a new algorithm for focusing tissue's scanning in real time with low computational cost and high performance. According to the results shown above, the two factors used to determine whether or not doing auto-focus are highly relevant, they explain the behavior of the focus based on the data of two adjacent images and groups of six ones. There is not a particular restriction in the use of the algorithm, the implementation is quite easy, and the precision is first class.

Acknowledgements

We appreciate the Center for Teaching Excellence CEDU Uninorte because of they bring spaces for the development of meetings and facilitate the creation of the interdisciplinary group. Also, We appreciate Dr. Maria Jose Herrera for her time to evaluate and propose consideration for optimizing the BIEMO method as well as Andrés Deulofeu, Miguel Padilla and Miguel Segrera for their constant support in the development of this article.

References

1. Hossain MS, Nakamura T, Kimura F, Yagi Y, Yamaguchi M. Practical image quality evaluation for whole slide imaging scanner. In Biomedical Imaging and Sensing Conference 2018 Apr 24 (Vol. 10711, p. 107111S). International Society for Optics and Photonics.
2. Andrew J Evans et al. "Implementation of whole slide imaging for clinical purposes: Issues to consider from the perspective of early adopters". In: Archives of pathology & laboratory medicine 141.7 (2017), pp. 944–959.
3. Vosoughi A, Smith PT, Zeitouni JA, Sodeman GM, Jorda M, Gomez-Fernandez C, Garcia-Buitrago M, Petito CK, Chapman JR, Campuzano-Zuluaga G, Rosenberg AE. Frozen section evaluation via dynamic real-time non-robotic Telepathology system in a university Cancer center by resident/faculty cooperation team. Human pathology. 2018 May 1.
4. Tabata K, Mori I, Sasaki T, Itoh T, Shiraishi T, Yoshimi N, Maeda I, Harada O, Taniyama K, Taniyama D, Watanabe M. Whole-slide imaging at

primary pathological diagnosis: Validation of whole-slide imaging-based primary pathological diagnosis at twelve Japanese academic institutes. Pathology international. 2017 Nov 1;67(11):547-54.

5. Farahani N, Liu Z, Jutt D, Fine JL. Pathologists' Computer-Assisted Diagnosis: A Mock-up of a Prototype Information System to Facilitate Automation of Pathology Sign-out. Archives of pathology & laboratory medicine. 2017 Jul 7;141(10):1413-20.
6. Farahani N, Liu Z, Jutt D, Fine JL. Pathologists' Computer-Assisted Diagnosis: A Mock-up of a Prototype Information System to Facilitate Automation of Pathology Sign-out. Archives of pathology & laboratory medicine. 2017 Jul 7;141(10):1413-20.
7. Lu Q, Liu G, Xiao C, Hu C, Zhang S, Xu RX, Chu K, Xu Q, Smith ZJ. A modular, open-source, slide-scanning microscope for diagnostic applications in resource-constrained settings. PloS one. 2018 Mar 15;13(3):e0194063.
8. Chantziantoniou N, Mukherjee M, Donnelly AD, Pantanowitz L, Austin RM. Digital Applications in Cytopathology: Problems, Rationalizations, and Alternative Approaches. Acta cytologica. 2018;62:68-76.
9. Farahani N, Liu Z, Jutt D, Fine JL. Pathologists' Computer-Assisted Diagnosis: A Mock-up of a Prototype Information System to Facilitate Automation of Pathology Sign-out. Archives of pathology & laboratory medicine. 2017 Jul 7;141(10):1413-20.
10. Jafarian AH, Tasbandi A, Mohamadian Roshan N. Evaluation of photoshop based image analysis in cytologic diagnosis of pleural fluid in comparison with conventional modalities. Diagnostic cytopathology. 2018 Apr 19.
11. Platiša L, Van Brantegem L, Kumcu A, Ducatelle R, Philips W. Influence of study design on digital pathology image quality evaluation: the need to define a clinical task. Journal of Medical Imaging. 2017 Jun;4(2):021108
12. Bueno G, Déniz O, Fernández-Carrobles MD, Vállez N, Salido J. An automated system for whole microscopic image acquisition and analysis. Microscopy research and technique. 2014 Sep 1;77(9):697-713.
13. Chan CC, Huang SK, Chen HH. Enhancement of phase detection for autofocus. InImage Processing (ICIP), 2017 IEEE International Conference on 2017 Sep 17 (pp. 41-45). IEEE.
14. Wang Y, Feng H, Xu Z, Li Q, Chen Y, Cen M. Fast auto-focus scheme based on optical defocus fitting model. Journal of Modern Optics. 2018 Apr 16;65(7):858-68.
15. Fu G, Cao Y, Lu M. A fast auto-focusing method of microscopic imaging based on an improved MCS algorithm. Journal of Innovative Optical Health Sciences. 2015 Sep;8(05):1550020.
16. Zhang C, Hou G, Zhang Z, Sun Z, Tan T. Efficient auto-refocusing for light field camera. Pattern Recognition. 2018 Sep 1;81:176-89.
17. Wang Y, Feng H, Xu Z, Li Q, Chen Y, Cen M. Fast auto-focus scheme based on optical defocus fitting model. Journal of Modern Optics. 2018 Apr 16;65(7):858-68.
18. Cabazos-Marín AR, Álvarez-Borrego J. Automatic focus and fusion image algorithm using nonlinear correlation: Image quality evaluation. Optik. 2018 Jul 1;164:224-42.
19. Nadeau JL, Bedrossian M, Lindensmith CA. Imaging technologies and strategies for detection of extant extraterrestrial microorganisms. Advances in Physics: X. 2018 Jan 1;3(1):1424032.
20. Choi S, Min SW. Depth extraction using depth of field imaging with tilted retroreflective structure. InConsumer Electronics (ICCE), 2018 IEEE International Conference on 2018 Jan 12 (pp. 1-3). IEEE.
21. Tian Y, Cui H, Pan Z, Liu J, Yang S, Liu L, Wang W, Li L. Improved three-dimensional reconstruction algorithm from a multifocus microscopic image sequence based on a nonsubsampled wavelet transform. Applied Optics. 2018 May 10;57(14):3864-72.
22. Li Q, Zhong J, Zhong S, Chen Z, Qi S, Zhang H. An self-focusing imaging method for leukocyte recognition. InImage and Signal Processing, BioMedical Engineering and Informatics (CISP-BMEI), 2017 10th International Congress on 2017 Oct 14 (pp. 1-5). IEEE.
23. Ishihara, T., Kitani, K. M., Asakawa, C., & Hirose, M. (2018, March). Deep Radio-Visual Localization. In Applications of Computer Vision (WACV), 2018 IEEE Winter Conference on (pp. 596-605). IEEE.

Transformation Thermodynamics with Arbitrarily Shaped Non-Conformal Objects and Coatings

Qingxiang Ji[1], *Guodong* Fang[1] and *Jun* Liang[2]

[1]Center for Composite Materials, Harbin Institute of Technology, Harbin, 150001, China
[2]Institute of Advanced Structure Technology, Beijing Institute of Technology, Beijing, 100081, China

Abstract. In this paper we apply transformation optics theory to thermodynamics and design thermal cloaks and concentrators with arbitrarily shaped non-conformal objects and coatings. Expressions of the required material parameters are derived analytically and then validated by numerical simulations. We apply this method to design a thermal cloak which can guide the heat flow around the inner domain without perturbation to external thermal fields. In this way, the object inside the inner domain is protected from the invasion of external heat fluxes. In contrast, a concentrator is designed to concentrate heat flows into a small region without disturbing outside temperature fields, which can considerably enhance the heat density in the designed domain. The proposed method extends the design flexibility in manipulating heat flux and will find wide applications in thermal protection systems, solar cells and so on.

1 Introduction

Transformation optics (TO) theory was first proposed by Pendry [1] and Leonhardt [2] in the field of achieving invisibility of electromagnetic waves. Since then TO attracts extensive interests and leads to considerable achievements in theoretical and experimental approaches extended to many fields, such as acoustics [3–5], magnetic field [6-8], elastic wave [9, 10] and matter waves [11, 12]. TO theory was also applied to thermodynamics since the diffusion equation retains its form under coordinate transformation [13–16]. Thermal cloaks, concentrators and other devices are designed based on coordinate transformations and then experimentally realized [17, 18]. These pioneer works theoretically demonstrate the possibilities for cloaking and focusing heat flux. Expressions of related diffusivity in a general coordinate system are also derived.

However, previous works are mainly focused on regular geometry such as cylindrical and spherical shapes. The geometric requirements impose great restrictions on practical applications and further research is needed. Guenneau [16] proposed two-dimensional thermal cloaks and concentrators with arbitrarily shaped objects and coatings. Yang [19] investigated a two-dimensional heat flux cloak with arbitrarily shaped objects and verified its theory by numeric results, but the work was about homotopy mapping that achieved same shaped objects and coatings.

In this paper, we propose three-dimensional heat flux devices with arbitrarily shaped non-conformal objects and coatings. General expressions of the transformation based parameters are derived. Then some finite element computations are performed, which demonstrates the designed properties of the cloaks and concentrators, consequently, verifies the validity of the proposed method. One of the practical strengths of this paper lies with its extension in design flexibility in manipulating heat fluxes.

2 Methodology

We consider a three dimensional diffusion problem without a heat source, of which the heat conduction equation can be written as

$$\nabla\bullet(\boldsymbol{\sigma}\nabla T)+c\rho\frac{\partial T}{\partial t}=0\,. \tag{1}$$

where $\boldsymbol{\sigma}$ is the heat conductivity and T is the temperature, c and ρ are the specific heat and material density, respectively.

Earlier works [10] in transformation thermodynamics has demonstrated that the conduction equation is invariant in its form under the coordinate transformation

$$\nabla'\bullet(\boldsymbol{\sigma}'\nabla'T')+c'\rho'\frac{\partial T'}{\partial t}=0\,, \tag{2}$$

when

$$\boldsymbol{\sigma}'=\frac{J\boldsymbol{\sigma}J^T}{\det(J)},\rho'c'=\frac{\rho c}{\det(J)}\,. \tag{3}$$

In Eq. (3), the superscript denotes that the parameters are expressed in the transformed space. J is the Jacobian transformation matrix with its elements derived by

$$J_{ij}=\left[\frac{\partial(x',y',z')}{\partial(x,y,z)}\right]. \tag{4}$$

Then a linear geometric transformation is defined as

$$\begin{cases} r' = \alpha(\theta)r + \beta(\theta) \\ \theta' = \theta \\ z' = z \end{cases} . \quad (5)$$

Eq. (5) denotes a radial stretch or compression of polar coordinates that squeezes space from a volume into a shell surrounding the object domain. The parameters $\alpha(\theta)$ and $\beta(\theta)$ indicates that the stretch or compression scale depends on the azimuthal angle when modeling domains of a smooth arbitrary shape.

Then the thermal conductivity and capacity parameters in transformed space are derived by

$$\boldsymbol{\sigma}' = \begin{bmatrix} \sigma_r \cdot \frac{r'-\beta}{r'} + \sigma_\theta \cdot \frac{\lambda^2}{r'(r'-\beta)} & \sigma_\theta \cdot \frac{\lambda}{r'-\beta} & 0 \\ \sigma_\theta \cdot \frac{\lambda}{r'-\beta} & \sigma_\theta \cdot \frac{r'}{r'-\beta} & 0 \\ 0 & 0 & \sigma_z \cdot \frac{r'-\beta}{\alpha^2 r'} \end{bmatrix} \quad (6)$$

$$\rho' c' = \frac{\rho c (r'-\beta)}{\alpha^2 r'}, \quad (7)$$

where $\lambda = \partial r' / \partial \theta$.

One should note that λ is not a constant value and varies in different transformations, which will be discussed in details in following sections. Clearly, the derivation of λ is the key step in cloaking and concentrating heat flux.

3 Results and discussion

3.1 An invisibility cloak

In this section we consider a two dimensional cloak based on the linear geometric transformation in Eq. (5). The inner boundary (boundary of the object) and outer boundary (boundary of the coating) are defined respectively by smooth closed curves $R_1(\theta)$ and $R_2(\theta)$ which depend upon the angular variable θ. We have parameters in this section expressed as follows:

$$\alpha = \frac{R_2(\theta) - R_1(\theta)}{R_2(\theta)}, \beta = R_1(\theta). \quad (8)$$

By Eq. (8), we will map the field within the object domain $r \le R_2(\theta)$ onto the coating domain $R_1(\theta) \le r' \le R_2(\theta)$. The region $r' \le R_1(\theta)$ defines the cloaking region while the coating area consists of a material with anisotropic conductivity given by

$$\lambda = \frac{1}{R_2(\theta)(R_2(\theta) - R_1(\theta))} \left\{ \frac{dR_1(\theta)}{d\theta} R_2(\theta)(R_2(\theta) - r') - \frac{dR_2(\theta)}{d\theta} R_1(\theta)(R_1(\theta) - r') \right\} . \quad (9)$$

It is easy to verify that if the inner boundary and outer boundary of the cloak are circular, namely $dR_1(\theta)/d\theta = dR_2(\theta)/d\theta = 0$, then λ vanishes.

To demonstrate the validity of the designed cloak, some finite element computations with COMSOL MULTIPHISICS are presented. We apply constant temperatures normalized to 1 and -1 respectively on the left side and right side while other sides are thermal insulated. The background material is defined as isotropic with unit conductivity. Different arbitrary geometries are used to describe the flexibility of the approach. Without loss of generality, the irregular boundaries of the cloak are defined on some different contour equations:

Case A:

$R_1(\theta) = 0.6 + 0.1\cos(\theta) + 0.05\sin(2\theta) - 0.1\sin(3\theta)$;

$R_2(\theta) = 1.2 + 0.1\sin(\theta) + 0.2\sin(3\theta)$

Case B:

$R_1(\theta) = 0.6; R_2(\theta) = 1.2 + 0.1\sin(\theta) + 0.2\sin(3\theta)$

Case C:

$R_1(\theta) = 0.6 + 0.1\cos(\theta) + 0.05\sin(2\theta) - 0.1\sin(3\theta)$;

$R_2(\theta) = 1.2$

Case D:

$R_1(\theta) = 0.6; R_2(\theta) = 1.2$

The simulated results for the temperature profile and the heat flux lines are shown in Fig. 1. The arrows display the pathway of the total heat flux while the white lines indicate the isothermal lines. It is seen that the heat flux pass around the object domain and eventually return undisturbed to their original trajectory as if they have passed through the object. Besides, the isothermal lines are smoothly bent around the inner region but perfectly excluded from these areas. In contrast, the temperature field inside the inner region is uniform and does not interact with the coating region. Therefore, the object inside the inner domain is protected from the invasion of the external heat flux.

It is seen that when the outer and inner boundaries are both circular the parameter λ vanishes. In this way, a perfect cylindrical thermal cloak is proposed, as is shown in Fig. 1 (d). The designed cloak manipulates heat flux around the inner domain, consequently, leads to thermal protection of the hidden objects.

Thus, a thermal cloak for arbitrarily shaped objects and coatings is achieved. Traditional method of thermal protection, such as building in thermally insulated materials to block part of the heat flows, is too simple and has single function. Besides, previous cloaks focusing on symmetry and regular shapes are also limited in practical applications because of their strict geometric restrictions. The method proposed in this paper makes it possible to design a cloak with arbitrarily different shaped objects and coatings which goes beyond geometric constraints and leads to more practical and extensive applications.

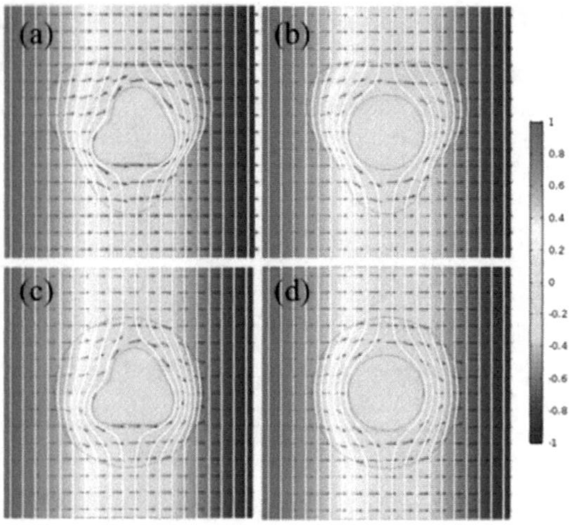

Fig. 1. Temperature field profiles of an invisibility cloak: (a) Case A; (b) Case B; (c) Case C; (d) Case D

3.2 A thermal concentrator

In this section we consider a two dimensional concentrator based on the linear geometric transformation in Eq. (5). The objects and coating area are defined by three embedded smooth closed curves $R_1(\theta) < R_2(\theta) < R_3(\theta)$, for $0 \le \theta \le 2\pi$. We define $\alpha = \frac{R_1(\theta)}{R_2(\theta)}$ and $\beta = 0$ for the domain $0 < r < R_2(\theta)$, thus mapping the field in the region $0 < r < R_2(\theta)$ onto the inner core $0 < r' < R_1(\theta')$. Similarly, by defining $\alpha = \frac{R_3(\theta) - R_1(\theta)}{R_3(\theta) - R_2(\theta)}$ and $\beta = R_3(\theta)\frac{R_1(\theta) - R_2(\theta)}{R_3(\theta) - R_2(\theta)}$ for the domain $R_2(\theta) < r < R_3(\theta)$, we can map the field in the region $R_2(\theta) < r < R_3(\theta)$ onto the coating area $R_1(\theta') < r' < R_3(\theta')$. The resulting thermic parameters are divided into two parts:

for the inner core $0 < r' < R_1(\theta')$,

$$\lambda = \frac{\frac{dR_1(\theta)}{d\theta} R_2(\theta) - \frac{dR_2(\theta)}{d\theta} R_1(\theta)}{R_1(\theta) R_2(\theta)} r', \quad (10)$$

for the coating area $R_1(\theta') < r' < R_3(\theta')$,

$$\lambda = -\frac{1}{(R_3(\theta) - R_1(\theta))(R_3(\theta) - R_2(\theta))} \cdot \left\{ \frac{dR_1(\theta)}{d\theta}(R_3(\theta) - R_2(\theta))(R_3(\theta) - r') - \frac{dR_2(\theta)}{d\theta}(R_3(\theta) - R_1(\theta))(R_3(\theta) - r') + \frac{dR_3(\theta)}{d\theta}(R_2(\theta) - R_1(\theta))(R_1(\theta) - r') \right\}. \quad (11)$$

It is obvious that if the concentrator is circular, namely $dR_1(\theta)/d\theta = dR_2(\theta)/d\theta = dR_3(\theta)/d\theta = 0$, then λ vanishes.

We report in Fig. 2 some finite element results for a concentrator when we apply constant temperature value normalized to 1 and -1 respectively on the left side and right side for a steady state and without heat source. Without loss of generality, the irregular boundaries of the cloak are defined on some different contour equations:

Case 1:

$$\begin{cases} R_1(\theta) = 0.6 + 0.1\cos(\theta) + 0.05\sin(2\theta) - 0.1\sin(3\theta) \\ R_3(\theta) = 1.2 + 0.1\sin(\theta) + 0.2\sin(3\theta) \end{cases}$$

Case 2:

$R_1(\theta) = 0.6;$

$R_3(\theta) = 1.2 + 0.1\sin(\theta) + 0.2\sin(3\theta)$

Case 3:

$R_1(\theta) = 0.6 + 0.1\cos(\theta) + 0.05\sin(2\theta) - 0.1\sin(3\theta);$

$R_3(\theta) = 1.2$

Case 4:

$R_1(\theta) = 0.6; R_3(\theta) = 1.2$

One should note that the boundary $R_2(\theta)$ is a virtual one and only serves the purpose of the geometric transformation. Therefore, we define a circular boundary with a radius of 0.85 for simplification, namely $R_2(\theta) = 0.85$ in the cases.

Temperature field profiles for different cases are shown in Fig. 2. One can see that the isothermal lines within the domain $R_1(\theta) \le r' \le R_2(\theta)$ are bent significantly towards the inner core, which increases the temperature gradient and thus the overall heat flow in the inner domain without affecting the exterior heat profile.

It is clear that if the concentrator is circular then the coefficient λ vanishes. A perfect cylindrical thermal concentrator is thus achieved, as is shown in Fig. 2 (d). The concentrator compresses the heat flux of the annulus into the small inner domain such that the energy density in that region is considerably enhanced while remaining uniform. It is found that this particular concentrator increases the heat flux gradient by 42% in the region $r' \le R_1(\theta)$.

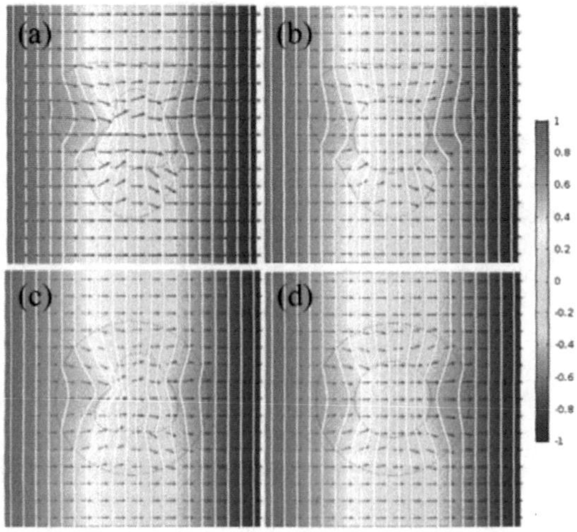

Fig. 2. Temperature field profiles for different cases of a concentrator: (a) Case 1; (b) Case 2; (c) Case 3; (d) Case 4.

4 Conclusions

In summary, this paper studied analytically and numerically the extension of transformation optics technique to the application of thermodynamics. We proposed three-dimensional heat flux cloaks and concentrators with arbitrarily shaped non-conformal objects and coatings. Such a cloak reduced the temperature inside the object without interfering with outside thermal field while a concentrator would significantly increase the heat density in a tiny region.

It is stressed that the proposed method is capable of controlling the diffusive heat current, including cloaking a given region and focusing heat flows into a domain of interest. The method we propose will extend the inherent design flexibility in transformation thermodynamics.

Acknowledgement

This work was supported by the National Natural Science Foundation of China (Grant Nos. 11732002, 11672089 and 11421091), the Fundamental Research Funds for the Central Universities (Grant No. HIT.NSRIF.2017017), Natural Science Foundation of Heilongjiang Province, China (Grant No. A2017003).

References

1. J.B. Pendry, D. Schurig, D.R. Smith, Science **312**, 1780 (2006)
2. U. Leonhardt, Science **312**, 1777 (2006)
3. M. Farhat, S. Guenneau, S. Enoch, A. Movchan, F. Zolla, A. Nicolet, New J. Phys. **10**, 115030 (2008).
4. S.A. Cummer, D. Schurig, New J. Phys. **9**, 45 (2009).
5. A.N. Norris, Proc. R. Soc. A **464**, 2411 (2008).
6. D. Schurig, J.J. Mock, B.J. Justice, S.A. Cummer, J.B. Pendry, A.F. Starr, D.R. Smith, Science **314**, 977 (2006).
7. F. Zolla, S. Guenneau, A. Nicolet, J.B. Pendry, Opt. Lett. **32**, 1069 (2007).
8. F. Magnus, B. Wood, J. Moore, K. Morrison, G. Perkins, J. Fyson, M. Wiltshire, D. Caplin, L. Cohen, J. B. Pendry, Nat. Mater. **7**, 295 (2008).
9. M. Brun, S. Guenneau, A.B. Movchan, Appl. Phys. Lett. **94**, 061903 (2009).
10. G.W. Milton, M. Briane, J.R. Willis, New J. Phys. **8**, 248 (2008).
11. A. Greenleaf, Y. Kurylev, M. Lassas, G. Uhlmann, New J. Phys. **10**, 115024 (2008).
12. S. Zhang, D.A. Genov, C. Sun, X. Zhang, Phys. Rev. Lett. **100**, 123002 (2008).
13. C.Z. Fan, Y. Gao, J.P. Huang, Appl. Phys. Lett. **92**, 251907 (2008).
14. X.Y. Shen, C.R. Jiang, Y. Li, J.P. Huang, Appl. Phys. Lett. **109**, 201906 (2016).
15. T. Chen, C.N. Weng, J.S. Chen, Appl. Phys. Lett. **93**, 114103 (2008).
16. S. Guenneau, C. Amra, D. Veynante, Opt. Express **20**, 8207 (2012).
17. S. Narayana, Y. Sato, Phys. Rev. Lett. **108**, 214303 (2012)
18. R. Schittny, M. Kadic, S. Guenneau, M. Wegener, Phys. Rev. Lett. **110**, 195901 (2013).
19. T. Yang, L. Huang, F. Chen, W. Xu, J. Phys. D: Appl. Phys. **46**, 305102 (2013).

Thermal Anomalies Detection Using Comparative Method for Small Earthquake

Pisanu Wongpornchai[1] and *Chanida* Suwanprasit [2]

[1]Department of Geological Sciences, Faculty of Science, Chiang Mai University, 50200 Thailand
[2]Department of Geography, Faculty of Social Science, Chiang Mai University, 50200 Thailand

Abstract. Thermal anomaly is one of the earthquake precursor in the earthquake preparatory phase. Remote sensing in thermal region has been employed based on the concept of stress accumulation in the active plate tectonics region, which may be transformed as temperature variation prior to earthquake. MODIS Land Surface Temperature has been commonly used to locate the thermal anomalies before the earthquake. Recently researches have been focusing on moderate or large magnitude earthquake events. In Thailand, small earthquake can severely damage the unprepared area. This study, the daily day- and night-time data of MODIS MOD11A1 product for 30 days before and 15 days after the earthquake on April 22, 2007, in Wiang Pa Pao District, Chiang Rai Province, Thailand, were processed and analysed to locate possibility of thermal anomalies. Thermal anomalies before and after the earthquakes were detected using the comparative method. The result found that the thermal anomaly temperature could be high up to 4.1 - 10.9 $^{\circ}$C which occurred in 21 - 22 days prior to the earthquake. Therefore, it may conclude that small earthquake can also release energy as the detectable thermal anomaly. However, more study about the relationship between thermal precursor and earthquake is needed to continue.

1 Introduction

Recently, earthquake prediction studies which are based on disastrous natural phenomenon correlated to the Earth's deformation, surface temperature growth, gas and aerosol exhalation, and electromagnetic disturbances in the ionosphere are more attended [1]. Thermal anomaly precursor has been attracted more in the earthquake prediction as it can be observed from the thermal remote sensors. Many reports depicted the appearence of the land surface temperature (LST) anomalies prior to strong earthquakes [2-7].

In many areas such as Middle Asia, Iran, China, Turkey, Japan, Kamchatka, India, Turkey, Italy, Greece, Spain and Thailand, the earthquake thermal anomalies have been studied on the earthquake with magnitude greater than 4.5 [1, 7]. However, the earthquake thermal anomalies associated with small magnitude earthquake was also interesting [8]. Many studies have found that the thermal remote sensors can detect the thermal anomalies from seismic activities [3, 4, 6-13].

Thailand is considered as a low seismicity area because its location is far from a plate boundary. However, the intraplate active faults and fractures are reported most in the northern and western parts of the country. Therefore, the small and moderate earthquakes which frequently occurred in the northern and western parts of Thailand were recorded.

Recently, many researches have been focusing on the moderate or strong magnitude earthquake events which can leading to severe unexpected damage on the natural environment, community, and life of peoples and animals. In Thailand, even though most earthquakes are small, the result from the earthquake can effect on the unprepared area. In this study, the surface thermal anomalies from a small earthquake with M4.5 on April 22, 2007 was analyzed from daily day- and night-time MODIS LST data during pre- and post-earthquake events to study the feasibility of using thermal remote sensing for detecting surface thermal anomalies.

2 Study area

In Thailand, active faults and fault zones are found in the northern, western and southern parts of the country. The study area is situated in an active major fault zone, called the Phayao Fault Zone (PFZ), northern part of Thailand. The PFZ is one of the 14 active major fault zones in Thailand. It passes through three provinces including Chiang Rai, Phayao and Lampang with 90 km long. It is trending in the NE to NW direction. On April 22, 2007, a magnitude 4.5 earthquake was detected and marked the epicenter at the PFZ at 19.40 °N latitude and 99.36 °E longitude, Wiang Pa Pao District, Chiang Rai Province, northern part of Thailand (Figure 1).

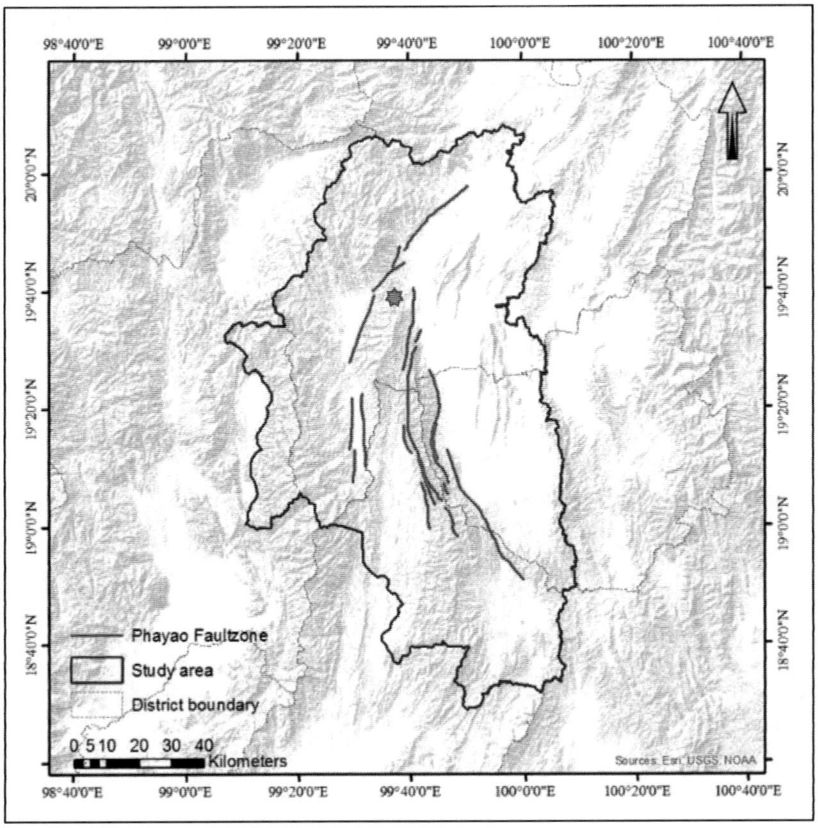

Fig. 1. The location map of the study area. The epicenter is represented by star.

3 Materials and methods

The data which were used in this study comprise of the land surface temperature (LST) satellite images. The daily LST satellite images (day- and night-time) over the study area from the thermal sensor with 1 km x 1 km resolution from Moderate Resolution Imaging Spectroradiometer (MODIS) on board Terra (MOD11A1) /Aqua (MYD11A1) Satellites were downloaded from the NASA's website. Since the earthquake was recorded on April 22, 2007, the image data set over a time period between March 23 – May 10, 2007 was used to analyze. The image data set over a time period between March 23 – May 10, 2008 was also downloaded as the reference background data set, since the earthquake was not detected for this year.

There were several steps of satellite image data processing. First, type of satellite remote sensing image was selected. Since the satellite images can be retrieved from many satellite systems, the factors those were considered for choosing the satellite images were the image resolution, data wavelength and satellite revisit time. This research concentrates on the large feature such as fault zone and the large area, so the suitable resolution at about 1 km was fine enough to use. Since the thermal anomaly appears in a very shot time, one revisit per day was considered. Second, preprocessing for satellite images was performed. This step was reselecting the suitable satellite images for doing processing since some images were covered by cloud and some image data were lost, so they could not be used. Third, the image mosaic and projection processes were performed since many satellite images were needed to mosaic for covering the huge study area and the difference in coordinate system of the satellite images was required to reproject.

Starting analysis process, the satellite images were converted into the surface temperature. The LST difference between the data in the earthquake year (2007) and the reference year (2008) at the same pixel location was calculated using the method which were purposed by Ouzounov and Freund [14]. Then, the comparative method was applied to detect the thermal anomalies during the pre- and post-earthquake as shown in the Equation (1).

$$\Delta LST = LST - LST_{bg(x/y)}, \qquad (1)$$

where

ΔLST is the thermal anomaly (°C),

LST is the average LST (°C),

$LST_{bg(x)}$ is the average LST before earthquake (°C),

$LST_{bg(y)}$ is the average LST after earthquake (°C).

Finally, the natural hot spot data from MODIS during the study period, hot spring locations, man-made hot spots and ground check survey were used for validate the surface thermal anomalies.

The surface thermal anomalies which were located at the same locations of natural hot spots, hot springs and man-made hot spots were eliminated from the surface thermal anomaly maps.

The ground check survey was the process in the field for searching for the evidences of relationship between surface thermal anomalies and faults. The surface thermal anomalies which were not related to the faults or fractures were also removed from the surface thermal anomaly maps.

4 Results

In the tropical country like Thailand, weather condition can affect the quality of the satellite images. During the time period between April 6 - April 20, 2007, the surface thermal anomaly was unable to analyze because most satellite images were covered by the clouds. Fortunately, some satellite images which were recorded before the earthquake event were still useful for the study.

On April 22, 2007, a small earthquake with magnitude of 4.5 was recorded at Wiang Pa Pao District, Chiang Rai Province, northern part of Thailand which is located in the Phayao Fault Zone.

The surface thermal anomaly from satellite images were divided into three groups as the pre-earthquake period (Mar 23 - April 21, 2007), earthquake event (April 22, 2007), and post-earthquake period (April 23 - April 30, 2007). The surface thermal anomalies of the day-time data from satellite images clearly show the surface thermal anomalies during the time period between April 1 - April 2, 2007 (20 - 21 days before the earthquake) with the maximum anomaly temperature of 10.9 °C on April 2, 2017, and during the time period between April 23 - April 24, 2007 (1 - 2 days after the earthquake) with the maximum anomaly temperature of 6.9 °C on April 24, 2007. (Figure 2).

The surface thermal anomalies of the night-time data from satellite images clearly show the surface thermal anomalies during the time period between March 31 - April 1, 2007 (21 - 22 days before the earthquake) with the maximum anomaly temperature of 4.2 °C on April 1, 2017, and on April 22, 2007 (the day of earthquake event) with the maximum anomaly temperature of 4.1 °C (Figure 3).

The main distribution of the surface thermal anomalies was found surrounding the epicenter. Most of the surface thermal anomalies appeared in the northeastern and central parts of the study area and mainly along the fault zone indicating the relationship between the surface thermal anomalies and the fault zone. However, some surface thermal anomalies can be found far from the main major fault zone, but they are related to the small faults and fractures.

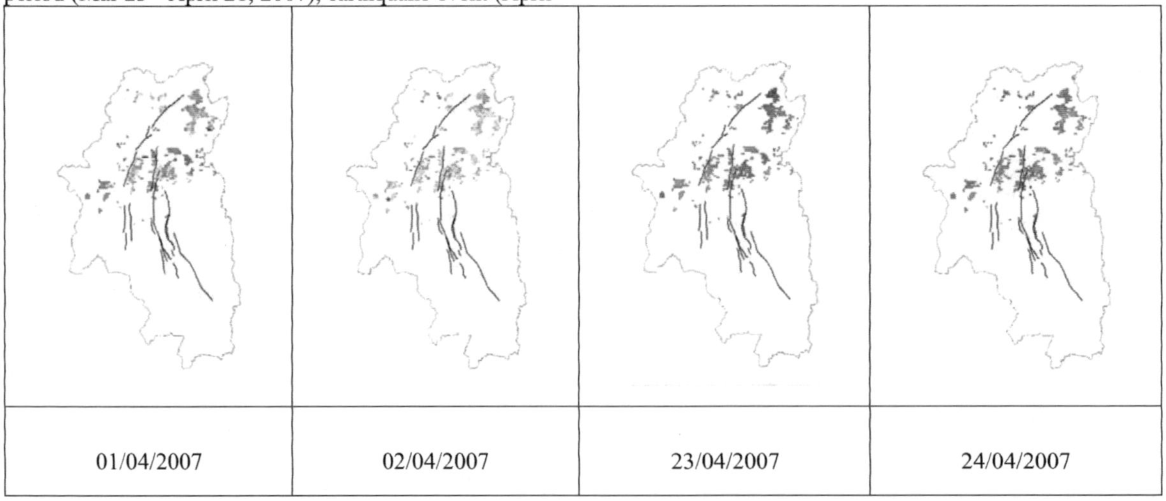

Fig. 2. The distribution of the surface thermal anomalies from the day-time data. The color scale is showm in Figure 3.

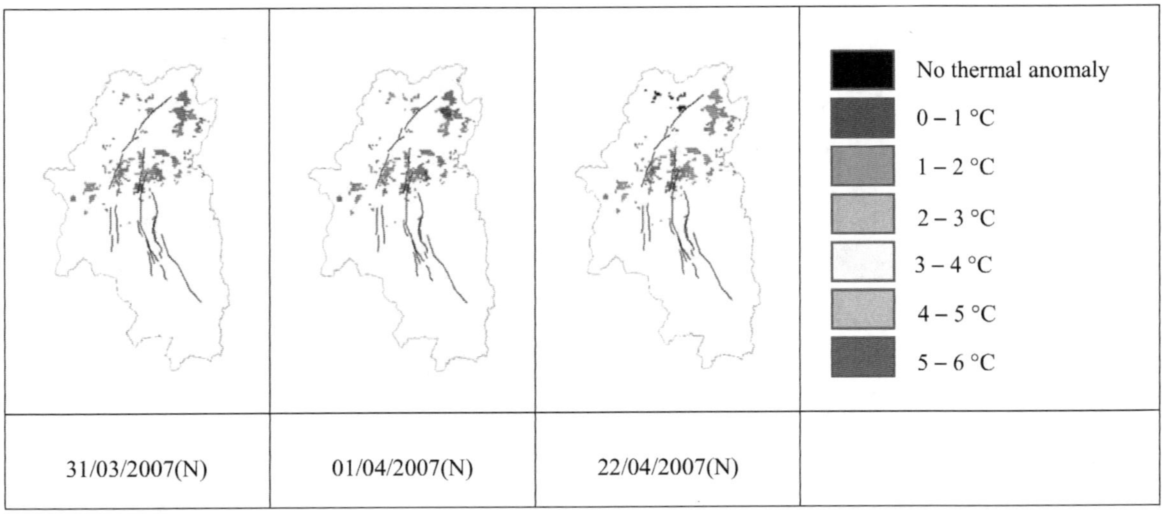

Fig. 3. The distribution of the surface thermal anomalies from the night-time data.

5 Conclusion

The earthquake is a complex geohazard event which several researchers have tried to discover the precursor before earthquake. Mostly, the study of precursor was in the moderate and strong earthquake events. This research used the advantage of the MODIS LST daily day- and night-time data to detect the pattern of surface thermal anomalies during pre- and post- small earthquake with M4.5 on April 22, 2007, in Wiang Pa Pao District, Chiang Rai province, northern part of Thailand.

Comparative method was applied for removing the background temperature using based year data. The validation process was done using natural and man-made hotspots, hot spring locations and ground survey data.

The results of thermal anomaly investigation from MODIS LST indicate: (1) the surface thermal anomalies appeared in 21-22 days before, about 1-2 days after and on the day of the earthquake event (2) the amplitude of the surface thermal anomaly is about 4.1 -10.9 °C. and (3) the surface thermal anomalies mainly appeared surrounding the epicentre and along the fault zone or local faults and fractures showing the relationship among the surface thermal anomalies, earthquake, and fault zone.

Since the optical satellite data in systematic and continuous monitoring of LST and surface thermal anomalies, cloud cover may act as a problem which effected to size of LST and thermal anomaly detection. Application on the other types of satellite image which will not be effected by the weather conditions should be developed for earthquake precursor investigation.

The small earthquake events in other cases should be tested to confirm this finding. This finding might be useful precursory sign for the observing fault activities and pre-earthquake events.

The local governments and Department of Dissaster Prevention and Mitigation will be benefit from this research.

6 Acknowledgments

This research was supported by The Thailand Research Fund (TRF) and Chiang Mai University, Thailand. The authors would like to acknowledge NASA for the MODIS LST data.

References

1. A.A. Tronin, Phys. Chem. Earth **31**,138 (2006)
2. B. Blackett, M.J. Wooster, B.D. Malamud, Geophys. Res. Lett. **38**, L15303 (2001)
3. S.K. Panda, C. Choudhury, A.K. Saraf, J.D. Das, Int. J. Rem. Sens. **28**, 4587 (2007)
4. M.R. Saradjian, M. Akhoondzadeh, Nat. Hazard Earth Sys. **11**,1099 (2011)
5. A.K. Saraf, V. Rawat, P. Banerjee, S. Choudhury, S. K. Panda, S. Dasgupta, J.D. Das, Nat. Hazards **47**,119 (2008)
6. M.A. Zoran, R.S. Savastru, D.M. Savastru, Open Geosciences **7**,606 (2015)
7. C. Wei, Y. Zhang, X. Guo, S. Hui, M. Qin, Y. Zhang, TSWJ **2013**, 208407 (2013)
8. P. Wongpornchai P, C. Suwanprasit, *International Multidisciplinary Scientific GeoConference SGEM 2017*, **21**, 903 (SGEM World Science, Albena, 2017)
9. M. Lisi, C. Filizzola, N. Genzano, R. Paciello, N. Pergola, V. Tramutoli, Phys. Chem. Earth **85-86**, 87 (2015)

10. M.A. Akhoondzadeh, Nat. Hazard Earth Sys. **13**, 1077 (2013)
11. W. Wu, L. Zou, X. Shen, S. Lu, N. Su, F. Kong, Y. Dong, J. Asian Earth Sci. **43**, 110 (2012)
12. W. Wu, L. Zou, X. Shen, S. Lu, N. Su, F. Kong, Int. J. Rem. Sens. **33**, 1850 (2012)
13. L. Piroddi L, G. Ranieri, IEEE J-STARS **5**, 307 (2012)
14. D. Ouzounov, F. Freund, Adv. Space Res. **33**, 268 (2004)

Symbolic-Numeric Computation of the Eigenvalues and Eigenfunctions of the Leaky Modes in a Regular Homogeneous Open Waveguide

Dmitriy Divakov [1], *Anastasiia* Tiutiunnik [1], and *Anton* Sevastianov [1]

[1]Peoples' Friendship University of Russia (RUDN University), 6 Miklukho-Maklaya St, Moscow, 117198, Russian Federation

Abstract. In this paper the algorithm of finding eigenvalues and eigenfunctions for the leaky modes in a three-layer planar dielectric waveguide is considered. The problem on the eigenmodes of open three-layer waveguides is formulated as the Sturm-Liouville problem with the corresponding boundary and asymptotic conditions. In the case of guided and radiation modes of open waveguides, the Sturm-Liouville problem is formulated for self-adjoint second-order operators on the axis and the corresponding eigenvalues are real quantities for dielectric media. The search for eigenvalues and eigenfunctions corresponding to the leaky modes involves a number of difficulties: the boundary conditions for the leaky modes are not self-adjoint, so that the eigenvalues can turn out to be complex quantities. The problem of finding eigenvalues and eigenfunctions will be associated with finding the complex roots of the nonlinear dispersion equation. In the present paper, an original scheme based on the method of finding the minimum of a function of several variables is used to find the eigenvalues. The paper describes the algorithm for searching for eigenvalues, the algorithm uses both symbolic transformations and numerical calculations. On the basis of the developed algorithm, the dispersion relation for the weakly flowing mode of a three-layer open waveguide was calculated in the Maple computer algebra system.

1 Introduction

The paper presents the formulation of the problem of numerical calculation of the leaky modes for planar three-layer waveguides.

The propagation of waveguide radiation in open waveguides is described by Maxwell's equations, boundary conditions at media interface, constitutive equations and asymptotic conditions [1-3].

The Maxwell's equations and the material equations in the Gaussian system of units have the following form [1]:

$$\operatorname{rot}\vec{E} = -\frac{1}{c}\frac{\partial \vec{B}}{\partial t}, \tag{1}$$

$$\operatorname{rot}\vec{H} = \frac{1}{c}\frac{\partial \vec{D}}{\partial t}, \tag{2}$$

$$\vec{D} = \varepsilon\vec{E}, \tag{3}$$

$$\vec{B} = \mu\vec{H}, \tag{4}$$

where $\vec{E}$, $\vec{H}$ - vectors of electric and magnetic field intensity; $\vec{D}$, $\vec{B}$ - vectors of electric and magnetic inductions; c - speed of electromagnetic wave propagation in vacuum; ε - dielectric permittivity, and μ - magnetic permeability of the medium.

At the interface of the dielectric waveguide layers, the boundary conditions imply the continuity of the tangential components of electromagnetic field [2,3]:

$$\left[\vec{E}^{\tau}\right]_{x=0,h} = \left[\vec{H}^{\tau}\right]_{x=0,h} = 0, \tag{5}$$

where h - thickness of the waveguide layer, $[f]_{s=a} = f(a-0) - f(a+0)$ - jump of the function f at the point $s=a$, and tangential components of the electromagnetic field $\vec{E}^{\tau}$ and $\vec{H}^{\tau}$ are defined as $\vec{E}^{\tau} = \vec{E}\times\vec{n}$, $\vec{H}^{\tau} = \vec{H}\times\vec{n}$ with the normal to the layers interface $\vec{n}$.

The asymptotic conditions for the leaky modes in planar waveguide structure have the following form:

$$\frac{\partial E_y}{\partial x} - ik_0\sqrt{n_c^2 - \beta^2}\,E_y \xrightarrow[x\to+\infty]{} 0, \tag{6}$$

$$\frac{\partial E_y}{\partial x} + ik_0\sqrt{n_s^2 - \beta^2}\,E_y \xrightarrow[x\to-\infty]{} 0, \tag{7}$$

where k_0 - wavenumber in vacuum, n_c - refractive index for the covering layer, n_s - refractive index of the substrate.

This work considers an open, regular, planar waveguide with three layers: a covering layer, a waveguide layer (film) and a substrate. The thickness of the covering layer and the substrate is semi-infinite, while the thickness of the waveguide layer is comparable to the wavelength. The refractive index is a piecewise constant function:

$$n(x) = \begin{cases} n_c, & x > h \\ n_f, & 0 \le x \le h \\ n_s, & x < 0, \end{cases} \quad (8)$$

where n_f - refractive index of the waveguide layer.

The radiation propagates along Oz axis. The geometry of such a structure is shown in figure 1.

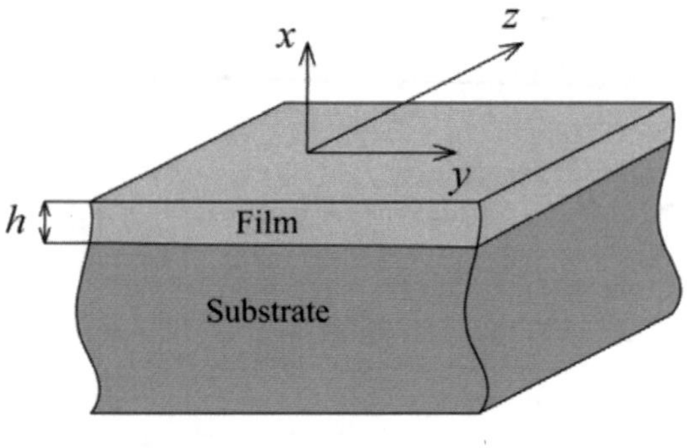

Fig. 1. Geometry of the dielectric waveguide structure

2 The problem formulation

For planar structures, time harmonic fields, under the assumption of the field invariance along the direction Oy, the Maxwell's equations are reduced to a pair of the Helmholtz equations: TE-mode and TM-mode; and two pairs of additional relations [1-6]:

$$\begin{cases} \left(\left(\dfrac{\partial^2}{\partial z^2}+\dfrac{\partial^2}{\partial x^2}\right)+k_0^2\varepsilon(x)\mu\right)E_y = 0 \\ H_x = -\dfrac{1}{ik_0\mu}\dfrac{\partial E_y}{\partial z} \\ H_z = \dfrac{1}{ik_0\mu}\dfrac{\partial E_y}{\partial x} \end{cases} \quad (9)$$

and

$$\begin{cases} \left(\dfrac{\partial^2}{\partial z^2}+\varepsilon(x)\dfrac{\partial}{\partial x}\left(\dfrac{1}{\varepsilon(x)}\dfrac{\partial}{\partial x}\right)+k_0^2\varepsilon(x)\mu\right)H_y = 0 \\ E_x = \dfrac{1}{ik_0\varepsilon(x)}\dfrac{\partial H_y}{\partial z} \\ E_z = -\dfrac{1}{ik_0\varepsilon(x)}\dfrac{\partial H_y}{\partial x} \end{cases} \quad (10)$$

In this paper we consider only the TE-mode. Applying the separation of variables method, we obtain the equation for the transverse and longitudinal parts [4-9]

$$\psi'' + k_0^2 n^2(x)\psi = k_0^2\beta^2\psi\,, \quad (11)$$

$$U'' = -k_0^2\beta^2 U\,. \quad (12)$$

In this paper we consider only the solution of the equation for the transverse part. We obtain the Sturm-Liouville problem:

$$\begin{cases} \psi'' + k_0^2 n^2(x)\psi = k_0^2\beta^2\psi \\ [\psi]\big|_{x=0,h} = [\psi']\big|_{x=0,h} = 0 \\ \psi' + ik_0\sqrt{n_s^2-\beta^2}\,\psi \xrightarrow[x\to-\infty]{} 0 \\ \psi' - ik_0\sqrt{n_c^2-\beta^2}\,\psi \xrightarrow[x\to+\infty]{} 0 \end{cases} \quad (13)$$

Taking into account the structure of the three-layer waveguide, the solution of the equation for the transverse part can be written in an analytical form in each of the regions, and for regions $x > h$ and $x < 0$ it is necessary to choose a solution that satisfies the asymptotic conditions. As a result, we can formulate a problem describing the eigenfunctions and eigenvalues corresponding to the leaky modes [10-12]:

$$\begin{cases} \psi'' + k_0^2 n_f^2\psi = k_0^2\beta^2\psi \\ \left(\psi' + ik_0\sqrt{n_s^2-\beta^2}\,\psi\right)\Big|_{x=0} = 0 \\ \left(\psi' - ik_0\sqrt{n_c^2-\beta^2}\,\psi\right)\Big|_{x=h} = 0 \end{cases} \quad (14)$$

3 Solution algorithm

In this paper we will consider an algorithm for finding the eigenvalues and functions for the leaky modes in a dielectric waveguide.

The equation for the transverse part of the waveguide layer has an analytical solution:

$$\psi(x) = A\,e^{ik_0\sqrt{n_f^2-\beta^2}\,x} + B\,e^{-ik_0\sqrt{n_f^2-\beta^2}\,x}, \quad (15)$$

where $\psi(x)$ corresponds to the solution for the waveguide layer. Substituting the solution form for the waveguide layer into the boundary conditions, we obtain a homogeneous system of linear algebraic equations (SLAE), represented as a product of the coefficient matrix **M** and vector of sought-for amplitude coefficients $\vec{A}$:

$$\mathbf{M}(\beta)\vec{A} = \vec{0}\,, \quad (16)$$

where $\vec{A} = \begin{pmatrix} A & B \end{pmatrix}^T$ and $\mathbf{M}(\beta)$ defined as

$$\mathbf{M}(\beta) = ik_0\begin{pmatrix} p_f + p_s & p_f - p_s \\ p_f\,e^{ik_0 p_f h} - p_c\,e^{ik_0 p_f h} & p_f\,e^{-ik_0 p_f h} + p_c\,e^{-ik_0 p_f h} \end{pmatrix}$$

and $p_\alpha = p_\alpha(\beta)$, $\alpha = c, f, s$ are presented below

$$p_f(\beta) = \sqrt{n_f^2-\beta^2}\,,\; p_s(\beta) = \sqrt{n_s^2-\beta^2}\,,\; p_c(\beta) = \sqrt{n_c^2-\beta^2}$$

To determine the sought-for values A and B it is necessary to solve a homogeneous SLAE.

In order for the system to have a nontrivial solution, it is necessary the determinant of the coefficient matrix of the system to be zero:

$$\det \mathbf{M}(\beta) = 0. \tag{17}$$

Since the matrix of coefficients has complex numbers, the solutions of the equation can also be complex numbers. Therefore, we move to the complex plane. Accordingly, it is necessary to find the acceptable values of the phase deceleration factor β : solutions of the equation and such complex numbers for which the imaginary and real parts of the determinant will be approximately equal to zero. To simplify the finding of β , we declare the function of squared module of the determinant, depending on two real numbers:

$$f(x, y) = \left|\det \mathbf{M}(x + iy)\right|^2. \tag{18}$$

Thus, the problem of finding β reduces to finding the minima of a function that depends on two real variables. Use the optimization method: pattern search. Since we know that the eigenvalues of the guided modes lie in the region from n_s to n_f , we assume that the real part of the eigenvalues of the leaky modes lies in the interval from 0 to n_s . So we construct on this interval a dense grid of initial approximations in order to apply the optimization method in each of them. The resulting problem can be formulated as follows

$$f(x, y) \xrightarrow[x \in [0; n_s],\, y > 0]{} \min. \tag{19}$$

Thus, having obtained all the values, we have to select only those that are not only minima of the function $f(x, y)$, but also are the zeros of this function. Naturally, they should not be repeated.

After finding all the eigenvalues, it is necessary to solve a homogeneous SLAE: find the sought-for amplitude coefficients. To do this, we use the modified Gauss method. This method allows us to find a nontrivial solution for matrices with the last zero row.

Each β has its own vector of amplitude coefficients:

$$\mathbf{M}(\beta_j)\vec{A}_j = \vec{0}. \tag{20}$$

This algorithm was implemented in the computer algebra system Maple [4, 5, 9, 10]. In the algorithm there is a function of the accuracy problem of calculations. The algorithm is available by the link: https://bitbucket.org/asdrevitskiy/bgcw-programs/.

4 Numeric results

Below we graphically represent the roots of the complex-valued function $\det \mathbf{M}(\beta)$ for a three-layer planar open waveguide characterized by $n_c = 1$, $n_f = 2.10$, $n_s = 1.47$, $h = 0.55\,\mu m$ and $\lambda = 0.55\,\mu m$.

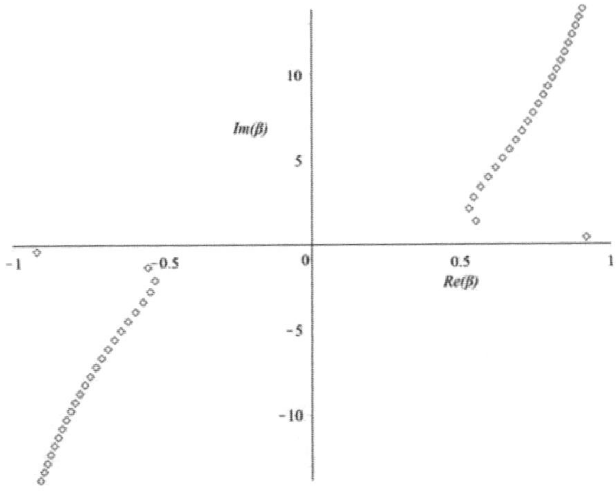

Fig. 2. Complex-valued coefficients β

As we can see from fig.2. there exists a complex eigenvalue with the smallest real part. The dispersion relation for leaky modes with the smallest real part of eigenvalues was calculated. For this purpose, the corresponding condition was written and integrated into the algorithm. The dispersion curves for real (Re) and imaginary (Im) parts of these eigenvalues are shown in figures 3, 4.

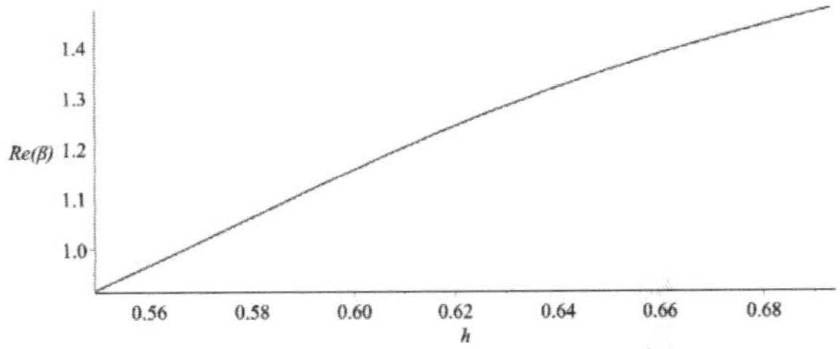

Fig. 3. Dispersion curve of the real (Re) part of β

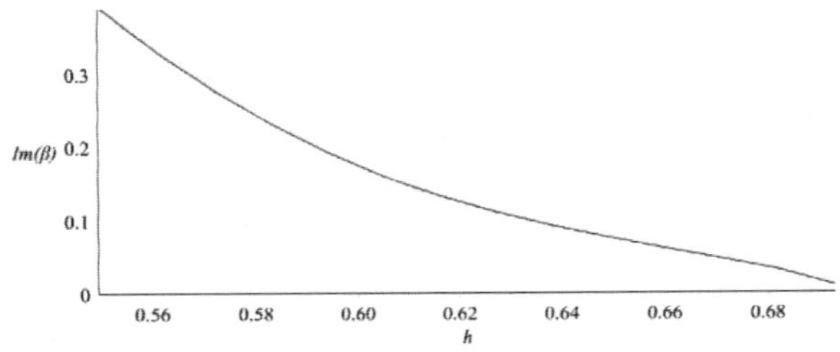

Fig. 4. Dispersion curve of the imaginary (Im) part of β

Also, for some of the eigenvalues (look for in Table 1) the eigenfunctions of the following modes were constructed, which are depicted in figures 5-8.

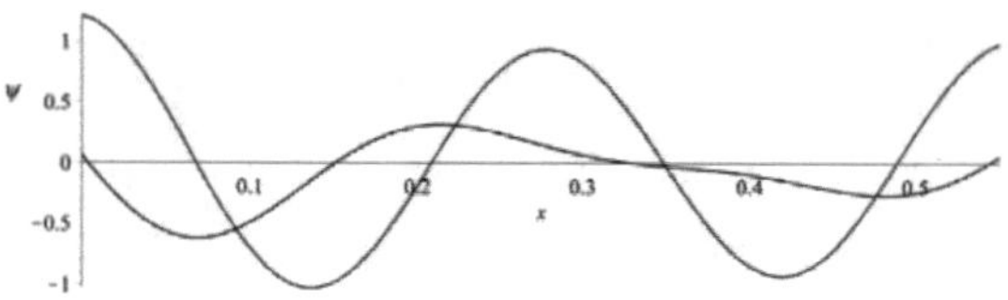

Fig. 5. Eigenfunction corresponding to eigenvalue $\beta = 0.918713 + 0.391051i$ (No.1 from Table 1)

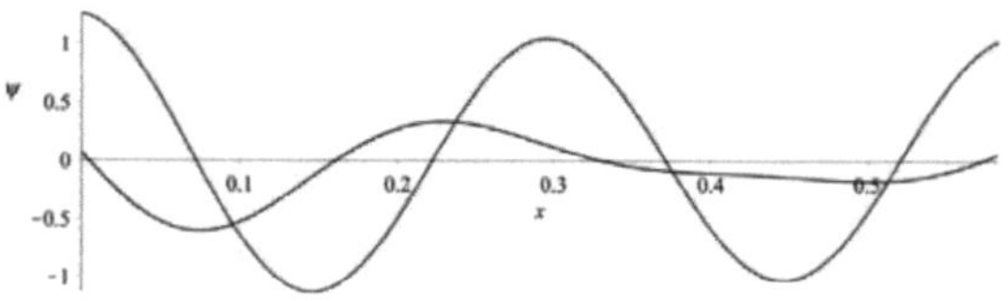

Fig. 6. Eigenfunction corresponding to eigenvalue $\beta = 1.076580 + 0.231889i$ (No.2 from Table 1)

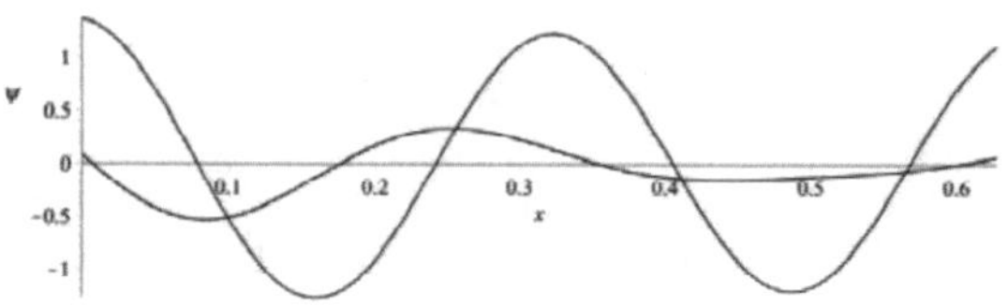

Fig. 7. Eigenfunction corresponding to eigenvalue $\beta = 1.271199 + 0.110653i$ (No.3 from Table 1)

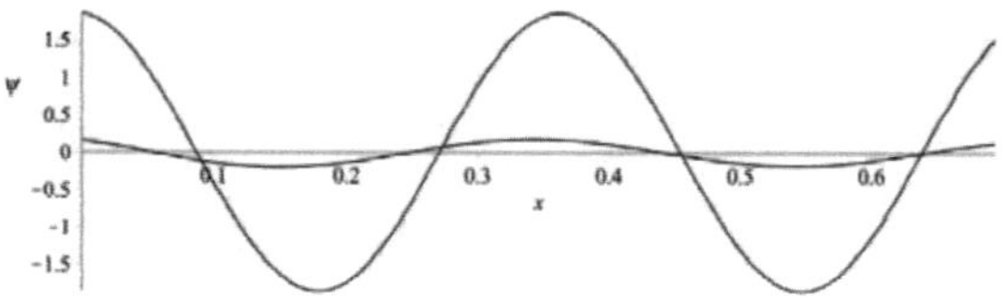

Fig. 8. Eigenfunction corresponding to eigenvalue $\beta = 1.473693 + 0.009583i$ (No.4 from Table 1)

Table 1. Eigenvalues

#	h	$\mathrm{Re}(\beta)$	$\mathrm{Im}(\beta)$
1	0.55	0.918713	0.391051
2	0.583	1.076580	0.231889
3	0.627	1.271199	0.110653
4	0.693	1.473693	0.009583

5 Conclusion

Within the framework of this paper, an algorithm for finding the eigenvalues and eigenfunctions of the leaky modes was presented. As a method of solution, the Hook-Jeeves method was chosen. Also, for the leaky modes, the dispersion dependence of the eigenvalues with the minimal imaginary part of the thickness of the waveguide layer was calculated.

The publication has been prepared with the support of the "RUDN University Program 5-100".

References

1. D. Marcuse, *Light Transmission Optics*, (Van Nostrand Reinhold, 1972).
2. M.J. Adams, *An Introduction to Optical Waveguides*, (John Wiley & Sons, New York, Wiley, 1981).
3. T. Tamir, *Guided-Wave Optoelectronics*, (Springer-Verlag Berlin Heidelberg, 1988).
4. D. Divakov, M. Malykh, L. Sevastianov, A. Sevastianov, E. Ayryan, *EPJ Web of Conferences*, **173**, 02007, (2018).
5. A.S. Drevitskiy, L.A. Sevastianov, D.V. Divakov, *CEUR Workshop Proceedings*, **1995**, 22-27, (2017).
6. A.A. Egorov, L.A. Sevast'yanov, A.L. Sevast'yanov, *Quantum Electronics*, **44**, 2, 167-173, (2014).
7. M.D. Malykh, L.A. Sevastianov, A.A. Tiutiunnik, N.E. Nikolaev, *Journal of Electromagnetic Waves and Applications*, **32**, 7, 886-898, (2018).
8. M. Malykh, L. Sevastianov, A. Tyutyunnik, N. Nikolaev, *EPJ Web of Conferences*, **173**, 02014, (2018).
9. L.A. Sevastianov, A.A. Egorov, A.L. Sevastyanov, *Physics of Atomic Nuclei*, **76**, 2, 224-239, (2013).
10. F. Emami, F. Hoseini, *International Journal of Fundamental Physical Sciences*, **3**, 2, 11-15, (2013).
11. S.K. Raghuwanshi, V. Palodiya, *Sensor Letters*, **13**, 9, 735-742, (2015).
12. A.B. Sotsky, L.M. Steingart, J.H. Jackson, P.Ya. Chudakovskii, L.I. Sotskaya, *Technical Physics*, **58**, 11, 1651-1660, (2013).

Algorithm for Numerical Solution of Diffraction Problem on the Joint of Two Open Three-Layer Waveguides

Dmitriy Divakov [1], *Anastasiia* Tiutiunnik [1], and *Anton* Sevastianov [1]

[1]Peoples' Friendship University of Russia (RUDN University), 6 Miklukho-Maklaya St, Moscow, 117198, Russian Federation

Abstract. This paper describes the algorithm for the numerical solution of the diffraction problem of waveguide modes at the joint of two open planar waveguides. For planar structures under consideration, we can formulate a scalar diffraction problem, which is a boundary value problem for the Helmholtz equation with a variable coefficient in two-dimensional space. The eigenmode problem for an open three-layer waveguide is the Sturm-Liouville problem for a second-order operator with piecewise constant potential on the axis, where the potential is proportional to the refractive index. The described problem is singular and has a mixed spectrum and therefore the Galerkin method can not be used in this definition. One way to adapt the Galerkin method for the problem solution is to artificially limit the area, which is equivalent to placing the open waveguide in question in a hollow closed waveguide whose boundaries are remote from the real boundaries of the waveguide layer of the open waveguide. Thus, we obtain a diffraction problem on a finite interval and with a discrete spectrum, which can be solved by the projection method, as done in this paper.

1 Introduction

The paper presents the definition of the diffraction problem for waveguide modes at the joint of two open planar three-layer waveguides. The waveguide joint is the simplest model of waveguide transition, and therefore the description of the diffraction problem at the interface is similar to the definition of the problem for a waveguide transition [1-3]. The classical method for solving diffraction problems in closed waveguides is the so-called incomplete Galerkin method [4-7]. The incomplete Galerkin method is a projection method for solving waveguide problems, which uses the Kantorovich decomposition to represent the approximate solution [8]. For description of the problem, Sveshnikov's partial radiation conditions are used [9, 10].

The propagation of radiation in open waveguides is described by Maxwell's equations, boundary conditions, material equations, and asymptotic conditions.

The Maxwell's equations and the material equations in the Gaussian system of units have the following form [1-3, 11]:

$$\operatorname{rot}\vec{E} = -\frac{1}{c}\frac{\partial \vec{B}}{\partial t},\ \operatorname{rot}\vec{H} = \frac{1}{c}\frac{\partial \vec{D}}{\partial t}, \qquad (1)$$

$$\vec{D} = \varepsilon \vec{E}, \vec{B} = \mu \vec{H}, \qquad (2)$$

where $\vec{E}$, $\vec{H}$ are vectors of electric and magnetic field intensities; $\vec{D}$, $\vec{B}$ are vectors of electric and magnetic inductions; c is speed of light in vacuum; ε is dielectric permittivity, and μ is magnetic permeability of the medium.

The boundary conditions on the boundaries of dielectric waveguide layers, imply the continuity of the tangential components of the electromagnetic field [1-3]:

$$\left[\vec{E}^{\tau}\right]_{x=0,h_a} = \left[\vec{H}^{\tau}\right]_{x=0,h_a} = 0 \qquad (3)$$

where $\left[f\right]_{s=a} = f(a-0) - f(a+0)$ is a jump of the function f at the point $s=a$, and tangential components of the electromagnetic field $\vec{E}^{\tau}$ and $\vec{H}^{\tau}$ are defined as $\vec{E}^{\tau} = \vec{E}\times\vec{n}$, $\vec{H}^{\tau} = \vec{H}\times\vec{n}$, where $\vec{n}$ is a normal to the layers interface.

The asymptotic conditions at an infinite distance from the upper boundary of the waveguide layer and, respectively, the lower boundary, are described assuming the field are bounded [1-3]:

$$\left|\vec{E}\right|_{x\to\pm\infty} \le C_E,\ \left|\vec{H}\right|_{x\to\pm\infty} \le C_H \qquad (4)$$

where C_E and C_H are some non-negative constants, and for guided waveguide modes $C_E = C_H = 0$.

2 Problem definition for TE-modes

We will consider open three-layer waveguides of different cross-section and with different refractive indices of the waveguide layers. Waveguides are joined at $z=0$. Each of the three-layer waveguides occupies the entire space, the waveguide layer of the left waveguide is characterized by thickness h_1 and refractive index $n_f^{(1)}$, right one is described by h_2 and

$n_f^{(2)}$. Covering layer ($x > h_{1,2}$) is described by refractive index n_c. The substrate ($x < 0$) is described by refractive index n_s and for both waveguides has semi-infinite thickness. The structure described is illustrated on Fig. 1.

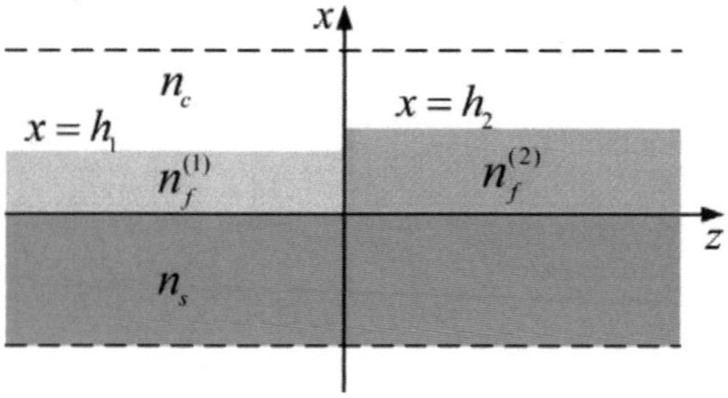

Fig. 1. Structure of two waveguides joint.

Refractive index $n(x,z)$ of the illustrated structure is defined as follows:

$$n_1(x) = \begin{cases} n_c, & x > h_1 \\ n_f^{(1)}, 0 < x < h_1 \\ n_s, & x < 0 \end{cases}, \quad n_2(x) = \begin{cases} n_c, & x > h_2 \\ n_f^{(2)}, 0 < x < h_2 \\ n_s, & x < 0 \end{cases}. \quad (5)$$

where $n(x,z)\big|_{z<0} = n_1(x)$ and $n(x,z)\big|_{z>0} = n_2(x)$.

The propagation of radiation in open waveguides is described by Maxwell's equations, boundary conditions, and material equations [1-3]. The geometry of planar structures does not change along the axis *Oy* and for time-harmonic fields (that do not change along axis *Oy* as well) Maxwell's equations are reduced to two independent subsystems - a subsystem for TE-modes and a subsystem for TM-modes [1-3]. Each subsystem can be described by a single Helmholtz equation with respect to the leading component of the field and two additional relations [1-3]. Therefore, for the planar structures under consideration, we can describe scalar diffraction problem, which is a boundary value problem for the Helmholtz equation with a variable coefficient in two-dimensional space.

The components of time-harmonic fields depend on time as $e^{-i\omega t}$. For the fields that are constant along Oy, the relations $\partial E_\alpha / \partial y = 0$ and $\partial H_\alpha / \partial y = 0$ where $\alpha = x, y, z$, are fulfilled. With the assumptions described above, Maxwell's equations are represented as a subsystem for TE-modes [1-6, 9, 12]:

$$\begin{gathered} \frac{\partial H_x}{\partial z} - \frac{\partial H_z}{\partial x} = -ik_0 \varepsilon E_y, \\ H_x = -\frac{1}{ik_0 \mu} \frac{\partial E_y}{\partial z}, H_z = \frac{1}{ik_0 \mu} \frac{\partial E_y}{\partial x} \end{gathered} \quad (6)$$

and a second subsystem linking the components of the TM-modes:

$$\begin{gathered} \frac{\partial E_x}{\partial z} - \frac{\partial E_z}{\partial x} = ik_0 \mu H_y, \\ E_x = \frac{1}{ik_0 \varepsilon} \frac{\partial H_y}{\partial z}, E_z = -\frac{1}{ik_0 \varepsilon} \frac{\partial H_y}{\partial x} \end{gathered} \quad (7)$$

where k_0 is wave number in vacuum, ε is dielectric permittivity, and μ is magnetic permeability of the waveguide. The subsystem for TE-modes can be represented as the Helmholtz equation with respect to the leading component E_y and two relations describing H_x and H_z by the leading component. Thus, the diffraction problem for a planar structure can be defined as a scalar problem - a boundary value problem for the Helmholtz equation.

We will consider only the TE-mode, the description of the problem for the TM-mode is constructed in a similar way. For the planar structures under consideration, we can formulate a scalar diffraction problem, which is a boundary value problem for the Helmholtz equation with a variable coefficient in two-dimensional space. The subsystem for the TE-mode can be formulated as the Helmholtz equation [1-6, 9, 12]:

$$\left(\frac{\partial^2}{\partial x^2} + \frac{\partial^2}{\partial z^2} + k_0^2 n^2(x,z) \right) E_y = 0, \quad (8)$$

where $n^2(x,z)$ is the refractive index of the waveguide under consideration, and two additional relations

$$H_x = -\frac{1}{ik_0 \mu} \frac{\partial E_y}{\partial z}, \; H_z = \frac{1}{ik_0 \mu} \frac{\partial E_y}{\partial x}. \quad (9)$$

The left and right waveguides joining at $z = 0$, are three-layer regular waveguides, that is, for the left and right waveguides, the refractive indices in the equation are different functions, depending only on x. That's why we can split the variables for the left and right waveguides. At $z = 0$ the boundary conditions of the continuity of the tangential components of the field are applied. The main difficulty in the study of diffraction problems in open waveguides is the existence of a continuous spectrum for the Sturm-Liouville problem for the transverse part E_y after applying the method of separation of variables. In this paper, an artificial limitation of the area is used, which is equivalent to placing the open waveguide under consideration in a hollow closed waveguide which boundaries are remote from the real boundaries of the waveguide layer of the open waveguide. That is, we consider the problem not on the axis $-\infty < x < \infty$ but on the segment $-R \le x \le R$, where R is several times greater than the thickness of the waveguide layer. The Dirichlet conditions $E_y\big|_{x=\pm R} = 0$ are fulfilled on borders $x = \pm R$. In the described method the continuous spectrum of the initial problem on the axis is roughly approximated by the discrete spectrum of the problem on the interval. The obtained approximate problem with a purely discrete spectrum can be solved using the Galerkin method. The described approach are called “the Dirichlet box”.

By applying the separation of variables method to the Helmholtz equation (8) for the right waveguide and for the left waveguide, we can obtain a solution in the corresponding waveguide as a superposition of modes $C_j^{(\alpha)}\varphi_j^{(\alpha)}(x)e^{\pm ik_0\beta_j^{(\alpha)}z}$, where $C_j^{(\alpha)}$ determines the amplitude of the corresponding mode, the sign in the exponential factor determines the direction of propagation of the mode: the positive sign corresponds to modes propagating from $-\infty$ in the positive direction of the axis z, $\beta_j^{(\alpha)}$ and $\varphi_j^{(\alpha)}(x)$ are eigenvalues and eigenfunctions of Sturm-Liouville problems, corresponding the left ($\alpha=1$) and right ($\alpha=2$) waveguides accordingly [4-6, 9]:

$$\begin{cases} \varphi''+k_0^2n_\alpha^2(x)\varphi=k_0^2\beta^2\varphi, \\ [\varphi]_{x=0,h_\alpha}=[\varphi']_{x=0,h_\alpha}=0, \\ \varphi(\pm R_x)=0. \end{cases} \quad (10)$$

During solution of "the Dirichlet box", we consider the diffraction problem of the guided mode, which propagates in the left waveguide from $-\infty$ in the positive direction of axis z. In the case where the left waveguide has the same configuration as the right waveguide, the guided mode passes into the right waveguide without losses. If, however, the configuration of the right waveguide differs from the configuration of the left waveguide, then reflected waveguide modes can appear in the left waveguide - both guided and radiative. In the right waveguide, the transmitted waveguide modes can appear as well, guided and radiative. The field in the left waveguide ($z<0$) can be represented in the form of a decomposition [4-6, 9]:

$$E_y^{(1)}(x,z)=Ae^{ik_0\beta_1^{(1)}}\varphi_1^{(1)}(x)+\sum_{j=1}^{\infty}R_je^{-ik_0\beta_j^{(1)}}\varphi_j^{(1)}(x) \quad (11)$$

where A determines the amplitude of the first waveguide mode, which propagates in the left waveguide from $-\infty$ in the positive direction of the axis z.

Amplitude reflection coefficients R_j determine the amplitudes of the reflected waveguide modes. The field in the right waveguide ($z>0$) can be represented in the form of decomposition

$$E_y^{(2)}(x,z)=\sum_{j=1}^{\infty}T_je^{ik_0\beta_j^{(2)}}\varphi_j^{(2)}(x) \quad (12)$$

where T_j determine the amplitudes of the transmitted waveguide modes. At $z=0$ the boundary conditions are satisfied:

$$\begin{cases} E_y^{(1)}\big|_{z=0}=E_y^{(2)}\big|_{z=0}, \\ \dfrac{\partial E_y^{(1)}}{\partial z}\Big|_{z=0}=\dfrac{\partial E_y^{(2)}}{\partial z}\Big|_{z=0}. \end{cases} \quad (13)$$

3 Symbolic solution

The equation for the Sturm-Liouville problem (10) is considered in three domains where $n^2(x)$ remains constant:

$$\psi_c''+k_0^2(n_c^2-\beta^2)\psi_c=0, h<x\le R_x, \quad (14)$$

$$\psi_f''+k_0^2(n_f^2-\beta^2)\psi_f=0, 0<x<h, \quad (15)$$

$$\psi_s''+k_0^2(n_s^2-\beta^2)\psi_s=0, -R\le x<0. \quad (16)$$

We'll solve the problem for the case (14). The characteristic equation is:

$$\lambda^2+k_0^2(n_c^2-\beta^2)=0. \quad (17)$$

The solution of this equation depends on the value β, that we still have to find. It is reduced to three cases:

$$\begin{cases} nf>\beta>ns>nc, \\ nf>ns>\beta>nc, \\ nf>ns>nc>\beta \end{cases} \quad (18)$$

For the first case, we obtain the following values of λ for three layers of the waveguide:

$$\begin{aligned} \lambda_{1,2}&=\pm ik_0\sqrt{\beta^2-n_f^2}, \\ \lambda_{3,4}&=\pm k_0\sqrt{\beta^2-n_c^2}, \\ \lambda_{5,6}&=\pm k_0\sqrt{\beta^2-n_s^2} \end{aligned} \quad (19)$$

which correspond to the following values of the functions ϕ:

$$\varphi_c=A_ce^{k_0\sqrt{\beta^2-n_c^2}(x-h)}+B_ce^{-k_0\sqrt{\beta^2-n_c^2}(x-h)}, h<x\le R_x \quad (20)$$

$$\varphi_f=A_f\sin\left(k_0\sqrt{n_f^2-\beta^2}x\right)+B_f\cos\left(k_0\sqrt{n_f^2-\beta^2}x\right), \quad (21)$$
$0<x<h$,

$$\varphi_s=A_se^{k_0\sqrt{\beta^2-n_s^2}x}+B_se^{-k_0\sqrt{\beta^2-n_s^2}x}, -R_x\le x<0. \quad (22)$$

For these solutions, we have 6 variables. In order to obtain these variables we need to solve a system of linear algebraic equations, which we compose by substituting (14)-(16) into the boundary conditions of the problem for the corresponding layers. The resulting system of linear equations can be represented in the form of the following equation:

$$M(\beta)\times\vec{A}=\vec{0}. \quad (23)$$

Since the solution $\vec{A}=\vec{0}$ is trivial for us, we need to find solutions for $|M(\beta)|=0$.

Substituting the eigenvalues into (23) and solving this system by the Gauss method with the choice of the main element, we obtain a set of eigenfunctions for the waveguide.

This search for eigenvalues and eigenfunctions is performed for each of the waveguides. Now we can proceed to solving the diffraction problem at the joint of the two waveguides.

The boundary conditions for this problem determine the continuity of the solution at the joint of two waveguides:

$$\begin{cases} E_y^{(1)}\big|_{z=0}=E_y^{(2)}\big|_{z=0}, \\ \dfrac{\partial E_y^{(1)}}{\partial z}\Big|_{z=0}=\dfrac{\partial E_y^{(2)}}{\partial z}\Big|_{z=0}. \end{cases} \quad (24)$$

Substituting formulas (11)-(12) into the boundary conditions (24), we obtain a system of relations for finding the amplitude coefficients of the reflection and transmission modes:

$$\begin{cases} A\varphi_1^{(1)}(x)+\sum_{j=1}^{+\infty}R_j\varphi_j^{(1)}(x)-\sum_{j=1}^{+\infty}T_j\varphi_j^{(2)}(x)=0, \\ ik_0\beta_n^{(1)}A\varphi_1^{(1)}(x)-ik_0\sum_{j=1}^{+\infty}\beta_j^{(1)}R_j\varphi_j^{(1)}(x)- \\ -ik_0\sum_{j=1}^{+\infty}\beta_j^{(2)}T_j\varphi_j^{(2)}(x)=0. \end{cases} \quad (25)$$

From this infinite system we obtain by truncation a system of size $N1+N2$, where $N1$ is the number of eigenfunctions for the left waveguide, $N2$ - for the right one. For scalar products, orthogonality conditions are observed in this system:

$$\begin{cases} \int_{-Rx}^{Rx}\varphi_i^{(1)}\varphi_j^{(1)}dx=0, i\neq j, \\ \int_{-Rx}^{Rx}\varphi_i^{(2)}\varphi_j^{(2)}dx=0, i\neq j. \end{cases} \quad (26)$$

This system can be represented as

$$\begin{cases} I\times\vec{r}-M_1\vec{t}=-A\times\vec{e}_1, \\ M_2\times\vec{r}-D\vec{t}=-A\times\vec{e}_2, \end{cases} \quad (27)$$

where $\vec{r}$ and $\vec{t}$ are vectors of amplitude coefficients R_j and T_j, $M_1=\left[m_{i,j}^{(1)}\right]$, where $m_{i,j}^{(1)}=\int_{-Rx}^{Rx}\varphi_j^{(2)}\varphi_i^{(1)}dx$, $M_2=\left[m_{i,j}^{(2)}\right]$, $m_{i,j}^{(2)}=ik_0\beta_i^{(2)}\int_{-Rx}^{Rx}\varphi_j^{(1)}\varphi_i^{(2)}dx$, I is identity matrix of size $N1\times N1$, $D=diag\{d_i\}$, where $d_i=ik_0\beta_i^{(2)}$, $\vec{e}_1=\{e_{i1}\}$, where $e_{i1}=\int_{-Rx}^{Rx}\varphi_i^{(1)}\varphi_i^{(1)}$, $\vec{e}_2=\{e_{i2}\}$, where $e_{i2}=\int_{-Rx}^{Rx}\varphi_i^{(1)}\varphi_i^{(2)}$.

By solving this system, we obtain the amplitude coefficients of the reflected and transmitted modes. By studying these modes we will be able to understand the efficiency of energy transfer for this case of waveguide joint.

4 Numerical solution

Let's solve this problem in one of the examples. We have a joint of two waveguides with heights $h_1=1.5\lambda$ and $h_2=\alpha\lambda$, inside the "box" of height $R_x=15\lambda$ ($\lambda=0.55\,\mu m$). They have a common substrate and a cover layer, the refractive indices of which are equal to $n_c=1$ and $n_s=1.47$. The refractive index of the waveguide layer material of the first waveguide is $n_f^{(1)}=1.565$, of the second one it is also $n_f^{(2)}=1.565$. One waveguide mode with amplitude coefficient of $A=1$ is running towards the joint.

The solution of this system is calculated in the *Maple 17* computer algebra system [5,6,7,11,12]. First, we search for eigenvalues and eigenfunctions for each of the waveguides. We consider the numerical solution for the left waveguide. The matrix determinant $M(\beta)$ graph is shown in Fig. 2.

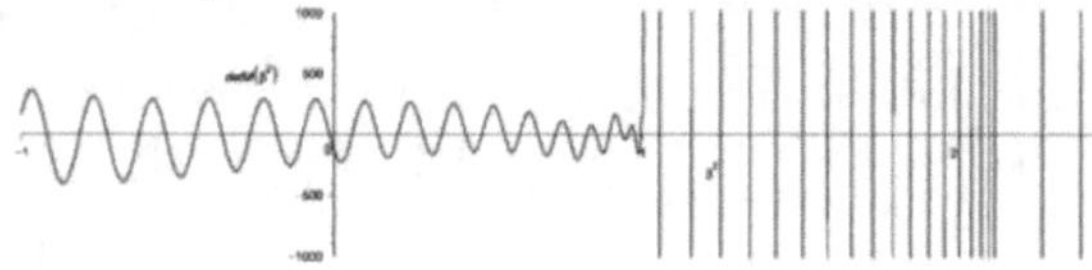

Fig. 2. Graph of determinant of $M(\beta)$.

By solving the equation $\det M(\beta)=0$ we obtain a set of eigenvalues $\beta_j^{(1)}$. Substituting the obtained values into the system and solving it by the Gauss method with the search for the main element, we obtain a set of coefficients $A_c, B_c, A_s, B_s, A_f, B_f$ for set of functions $\varphi_j^{(1)}$. By substituting them we obtain a set of eigenfunctions. The graphs of the obtained eigenfunctions are shown in Fig. 3–5.

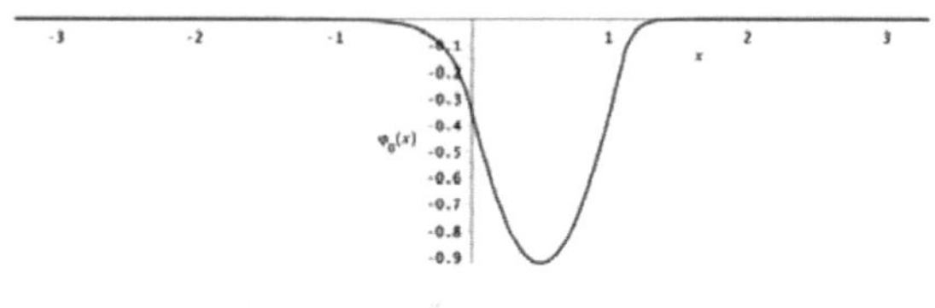

Fig. 3. Graph of $\varphi_1^{(1)}(x)$.

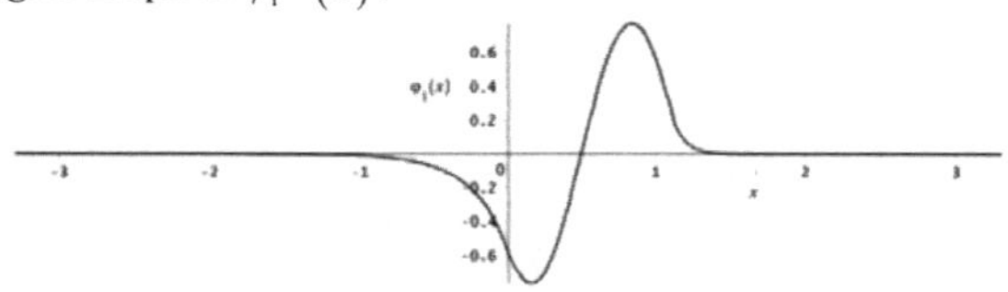

Fig. 4. Graph of $\varphi_2^{(1)}(x)$.

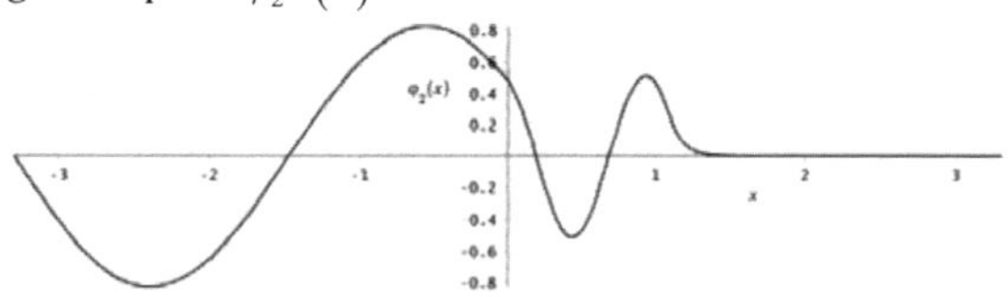

Fig. 5. Graph of $\varphi_3^{(1)}(x)$.

By performing the same operations for the second waveguide, we obtain a set of $\beta_j^{(2)}$ and $\varphi_j^{(2)}$. Now we can proceed to solving the diffraction problem at the joint of two waveguides.

After applying the projection relations of the Galerkin method, we obtained a system (27), the orthogonality conditions of the scalar products are applied for it. To solve the system and find the amplitude coefficients, we use the standard function *LinearSolve* of the *Maple 17* system. The resulting solution vector contains the amplitude coefficients of the reflected and transmitted waveguide modes.

We will investigate the variation of the transmission coefficients of the first two waveguide modes of the right waveguide as a function of the parameter α - see fig.6.

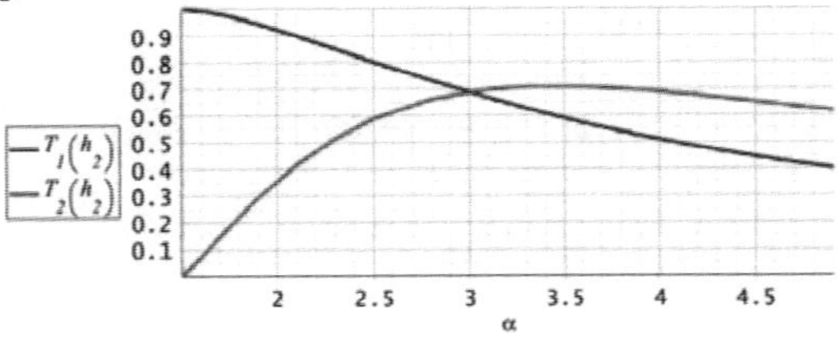

Fig. 6. Transmission coefficients of the first two waveguide modes.

As we can see from the fig.6. for $\alpha = 1.5$ - that is, when both waveguides have the same thickness - the transmission coefficient of the first mode is almost equal to 1, the transmission coefficient of the second mode is almost zero, since almost all of the energy is transferred by the first mode. As the parameter α is increased, the transmission coefficient of the first mode decreases, and the transmission coefficient of the second mode, on the contrary, increases. A further increase in the thickness of the right waveguide leads to the fact that the first mode incident on the waveguide joint excites simultaneously two directional modes, and for $\alpha = 3.5$ the second waveguide of the right waveguide mode transfers the main energy.

The accuracy of this solution can be determined by summing up the intensities of the transmitted and reflected modes and comparing them with the intensity of the incident mode:

$$\left| |A|^2 - \sum_{j=1}^{N1} |R_j|^2 - \sum_{j=1}^{N2} |T_j|^2 \right| = \delta \qquad (28)$$

In our case, the difference is $\delta = 0.0002$ (much less than 1%), that gives satisfactory accuracy of the solution.

5 Conclusion

In this paper we described the algorithm for solving the diffraction problem for waveguide modes at the joint of two planar three-layer waveguides. The proposed approach of the "Dirichlet box" makes it possible to approximately analyze the intensity distribution along the guided and radiative modes of the open waveguide for the diffraction problem of waveguide modes at the waveguide joint. We described the method, developed the symbolic-numerical algorithm based on it, and conducted the numerical experiment with the usage of presented algorithm.

The described algorithm makes it possible to study the energy transfer during the propagation of waveguide modes through junctions of different configurations and to numerically estimate the redistribution of energy between modes, which is important in waveguide matching problems.

The publication has been prepared with the support of the "RUDN University Program 5-100".

References

1. D. Marcuse, *Light Transmission Optics*, (Van Nostrand Reinhold, 1972).
2. M.J.Adams, *An Introduction to Optical Waveguides*, (John Wiley & Sons, New York, Wiley, 1981).
3. T.Tamir, *Guided-Wave Optoelectronics*, (Springer-Verlag Berlin Heidelberg, 1988).
4. A.G. Sveshnikov, *U.S.S.R. Comput. Math. Math. Phys.*, **3**, 1, 170–179, (1963).
5. D.V. Divakov, M.D. Malykh, L.A. Sevastianov, A.L. Sevastianov, E.A. Ayryan, *EPJ Web of Conferences*, **173**, 02007, (2018).
6. D.V. Divakov, L.A. Sevastianov, N.E. Nikolaev, *Journal of Physics: Conference Series*, **788**, 1, 012010, (2017).
7. A.A. Gusev, O. Chuluunbaatar, S.I. Vinitsky, et al., *Physics of Atomic Nuclei*, **75**, 10, 1210–1226, (2012).
8. L.V. Kantorovich, V.I. Krylov, *Approximate Methods of Higher Analysis*, (New-York, Wiley, 1964).
9. D.V. Divakov, L.A. Sevastianov, N.E. Nikolaev, *EPJ Web of Conferences*, **108**, 02020, (2016).
10. M.D. Malykh, *Journal of Mathematical Sciences*, **209**, 6, 935-952, (2015).
11. D.S. Kulyabov, A.V. Korolkova, L.A. Sevastianov, *Journal of Physics: Conference Series*, **788**, 1, 012025, (2017).
12. L.A. Sevastyanov, A.L. Sevastyanov, A.A. Tyutyunnik, *Lecture Notes in Computer Science*, **8660**, 419-431, (2014).

Effect of Annealing Process on the Mechanical Properties of X70 Pipeline Steel

Teng-wei Zhu[1], *Cheng-liang* Miao[1], *Zheng* Cheng[2], *Zhipeng* Wang[1], *Yang* Cui[1] and *Fei* Li[1]

[1]Shougang research institute of technology, Beijing 100043, China;
2Shougang Jingtang Steel Company, Hebei Caofeidian 063210, China

Abstract. The influence of the mechanical properties of X70 pipeline steel under different annealing temperature was studied. The corresponding microstructure was investigated by the Field Emission Scanning Electron Microscopy. The results showed that the yield strength and the tensile strength both experienced from rise to decline with the increase of annealing temperature. The grain sizes were coarse and a large amount of cementite precipitated due to preserving temperature above 550 °C, which induced matrix fragmentation and deteriorate the -10 °C DWTT Toughness. There were little changes on the microstructure and mechanical properties when the annealing temperature was under 500 °C.

1 Introduction

With the increasing gas pressure and more harsh service conditions, long-distance pipeline is gradually required to possess thick wall, large caliber, high strength and toughness [1, 2]. Generally, X70 and X80 are applied during most of the major projects all over the world with a thickness from 15mm to 22mm. They are efficient, economic and can last for several decades and have been recognized as the mostly utilized steels for pipeline construction in the petroleum industry [3]. It is reported that [4, 5], heat treatment is usually adopted to achieve better comprehensive performance during the pipe making majorly for high-grade. Annealing is influential on the mechanical properties especially for low temperature toughness [6]. Cleavage fracture splits in separations are usually detected on such material after tensile test, impact test and DWTT (Drop Weight Tear Test) [7]. Cleavage in separations is a typical kind of fracture for controlled rolled steels, the mechanism of which however lacks of research. In the present work, the effect of annealing process on the mechanical properties was studied and the formation mechanism of cleavage in separations after DWTT was investigated.

2. Experimental procedures

2.1 Materials and sample preparation

The material studied in this work is X70 pipeline steel with thickness 20mm. The composition is shown in Table 1.

Table 1. Chemical composition.

C/%	Si/%	Mn/%	Nb+V+Ti/%	Cr+Cu+Ni+Mo/%
0.06	0.2	1.70	≤1.50	≤0.60

The steel sample was taken from 1/4 position of the width and 45° to the rolling direction by flame cutting and was cut into rough samples with dimensions 420mm × 65mm × T used for tensile test, 45mm × 18mm × T for impact test and 300mm × 90mm × T for DWTT before heat treatment.

2.2 Experiment

Muffle furnace was applied for annealing treatment on prepared samples. The samples were heated in the furnace up to 400~650 °C. Preserved for 4 hours respectively before furnace cooling down to 150 °C and then taken out for air cooling. The annealing process is shown in Table 2.

Table 2. Annealing process.

Sample No.	Annealing process	
	Heating temperature	Preservation time
1#	Blank sample (Without heat treatment)	
2#	400 °C	4 hours
3#	500 °C	4 hours
4#	550 °C	4 hours
5#	600 °C	4 hours
6#	650 °C	4 hours

After annealing, the rough samples were machined into standard tensile samples (Gauge Length 50mm), impact test samples (10 × 10 × 55mm) and DWTT samples (Pressed Notch). Tensile test was conducted on Zwick100 tensile test machine. The impact test was performed on PSW750 full automatic impact test machine from Zwick at a speed of 20 mm/min. The linear velocity at the point of impact was around 5.5m/s. DWTT was performed on the JL40000 testing machine from China.

Using Metallography and Field Emission Scanning Microscope to investigate the microstructure after the samples were polished and eroded by 4% nital.

3. Results and discuss

3.1 Tensile test

The tensile properties with different annealing process are shown in Fig. 1. Rt0.5 and Rm goes through three stages with the increase of annealing temperature: (1) When the temperature is below 500°C, Rt0.5 increases with the annealing temperature very slightly while Rm stays the same; (2) After preserved under 550 °C for 4 hours, both Rt0.5 and Rm increase remarkably. Besides, the increase of Rt0.5 excesses Rm and as a result, the yield ratio increase rapidly; (3) When the temperature is beyond 550 °C, both Rt0.5 and Rm decrease while the yield ratio changes little.

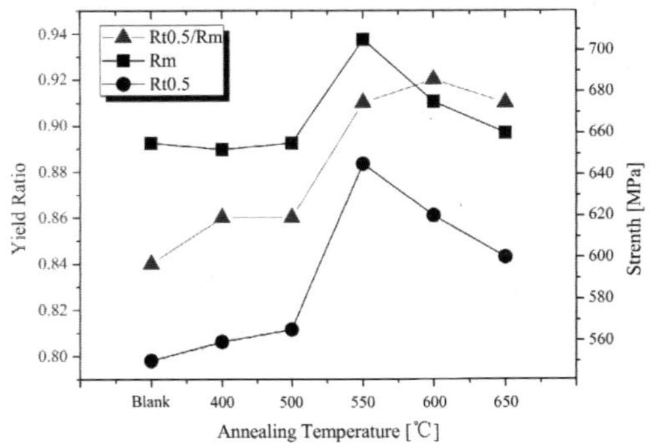

Fig. 1. Tensile test results of different annealing process.

The results of -20 ℃ impact test and -10 ℃ DWTT are shown in Fig. 2. The impact absorbed energy decreased with the increase of annealing temperature but the general level maintained in a fine condition. When the annealing up to 650 ℃, the absorbed energy can still be around 300 J. The result of DWTT exhibited the same trend. When the annealing temperature was under 500 °C, the shear ratio experienced little changes and stay around 100%. However, the shear ratio of 4#, 5# and 6# sample showed an obvious decrease. That is to say, the DWTT properties deteriorated when the annealing temperature was above 550 °C. The fracture morphologies are shown in Fig. 3. The fracture was mainly ductile when the annealing temperature is under 500 °C and became fragile with obvious separations when it was above 550 °C.

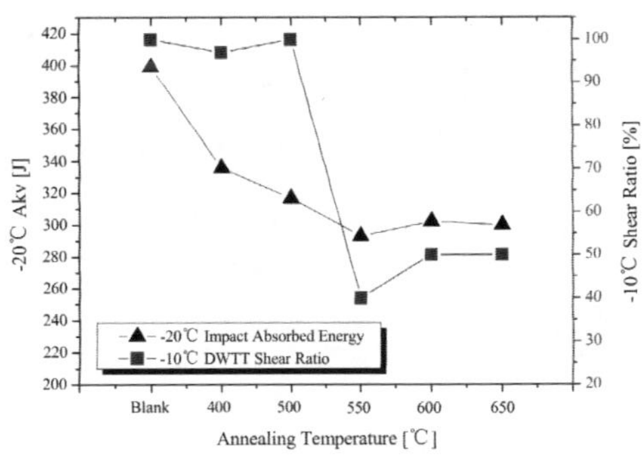

Fig. 2. Low temperature toughness with different annealing process.

https://doi.org/10.1051/matecconf/201818602001

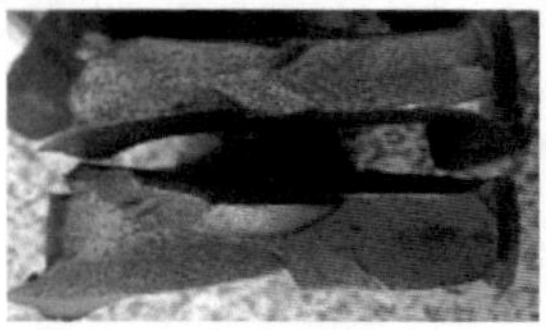

1# Blank

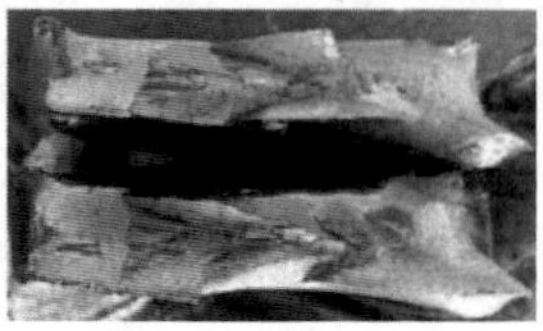

2#400 ℃

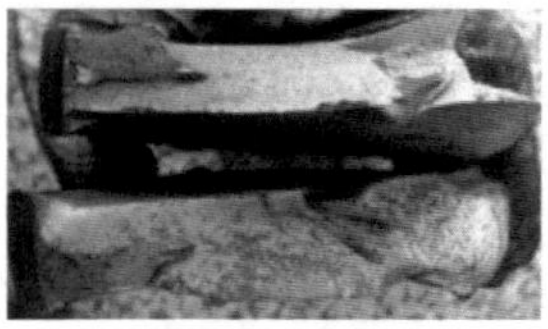

3# 500 ℃

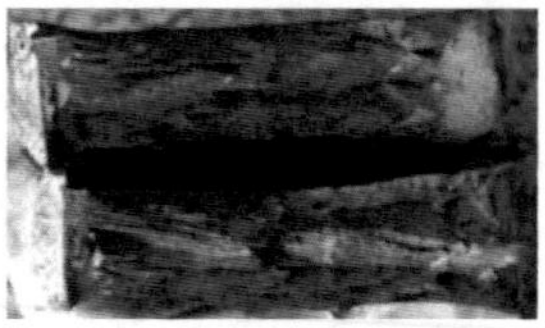

4# 550 ℃

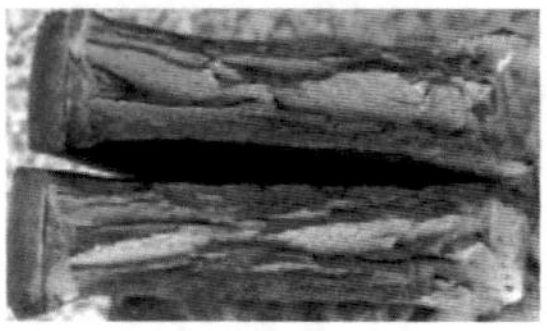

5# 600 ℃

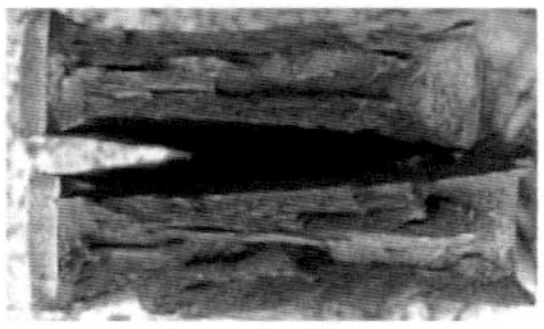

6# 650 ℃

Fig. 3. DWTT fracture morphology.

3.2 Microstructure observation

The microstructure was observed by optical microscope and the result is shown in Fig. **4**. The blank sample was dominated by acicular ferrite. There has been no significant change of microstructure when the annealing temperature was under 500 °C. However, the grain size increased from 5 μm to more than 10 μm when it was above 550 °C. The increase of grain size can become one of the main reasons of the decrease of low temperature impact properties.

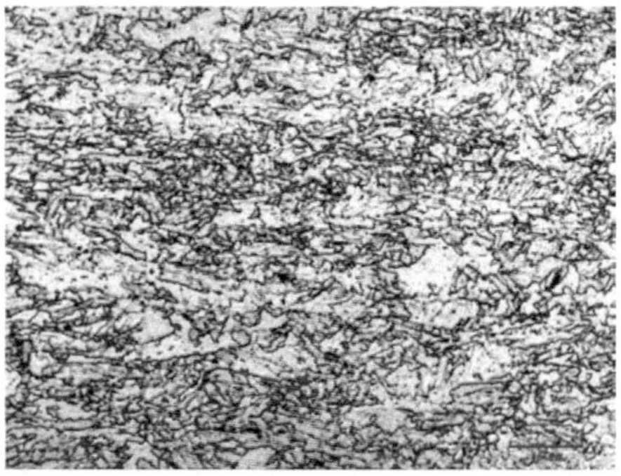

1# Blank

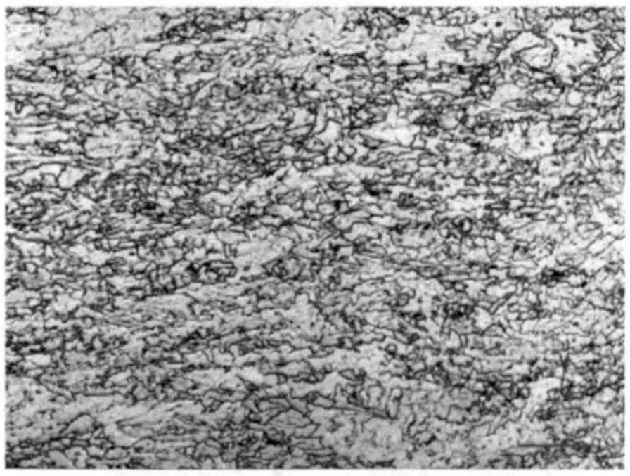

2# 400 ℃

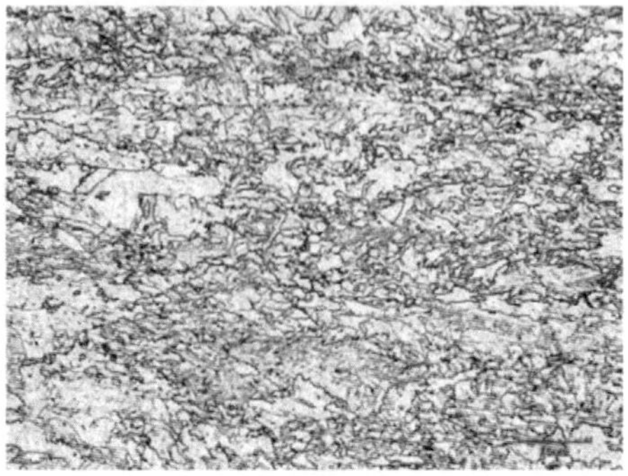

3# 500 ℃

4# 550 ℃

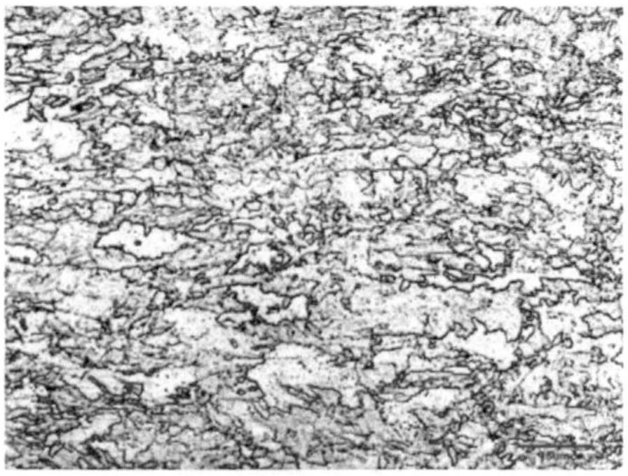

5# 600 ℃

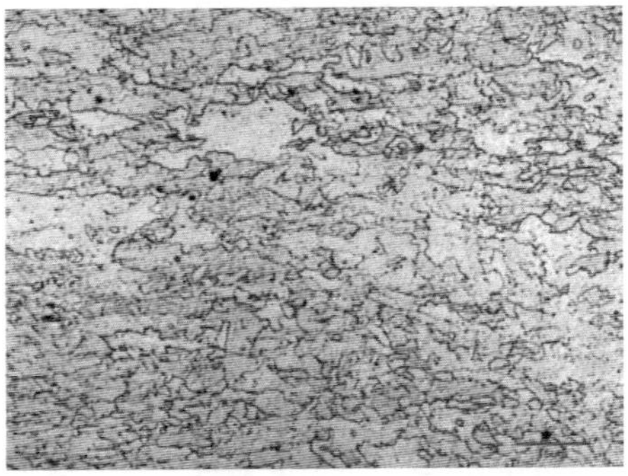

6# 650 ℃

Fig. 4. Microstructure morphologies under different annealing process.

3.3 Scanning Electron Microscopy Analysis

The SEM microstructure morphologies are shown in Fig. 5. Mass precipitation particles were detected in 4#, 5# and 6# sample. The particles were mainly distributed in grain boundaries judged from 4# and 5# while even precipitated inside the grain of 6#. That is to say, with the increase of annealing temperature, the precipitation particles increased both with their number and size. No obvious precipitation particles were observed when under 500 °C.

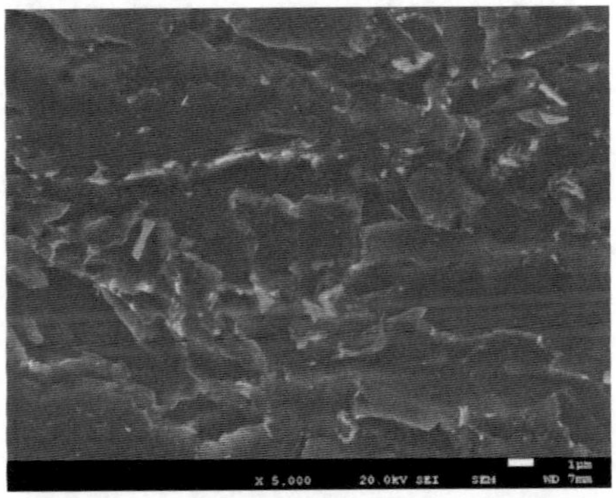

1# Blank

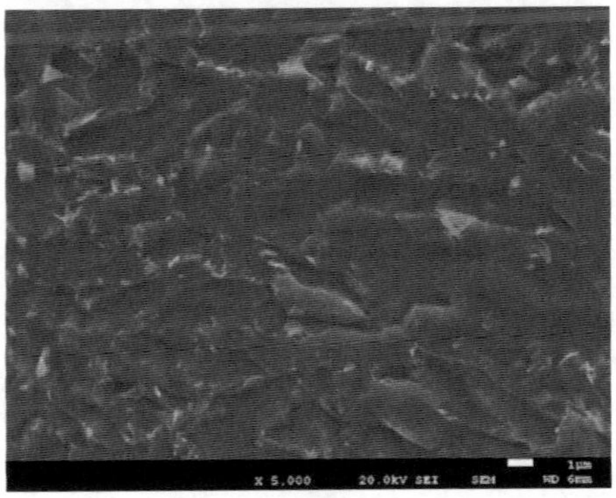

2# 400 ℃

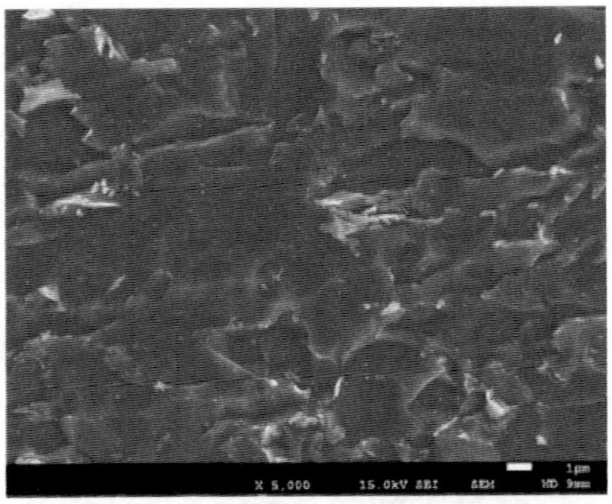

3# 500 ℃

4# 550 ℃

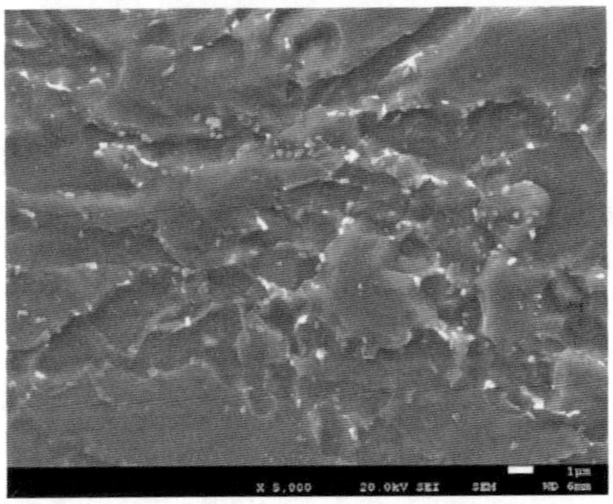

5# 600 ℃

6# 650 ℃

Fig. 5. SEM morphologies under different annealing process.

In order to identify the precipitation particles, the energy spectrum analysis was conducted; the result is shown as Fig. **6** and Table 3. The particles are mainly cementite. The cementite precipitation particles can destroy the uniformity of the matrix [8, 9]. When the material is faced to impact, the stress will concentrate on the junction of the precipitation particles and the matrix and finally lead to fracture.

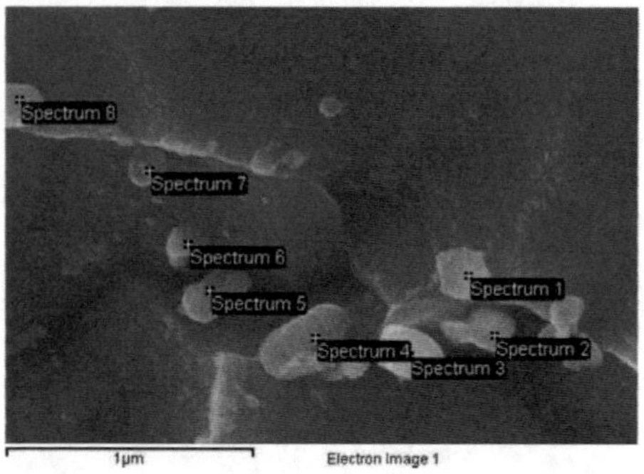

Fig. 6. Precipitation particles Energy Spectrum Analysis.

Table 3. Spectrum Results.

Spectrum	In stats.	C	Cr	Mn	Fe	Total
Spectrum 1	Yes	11.37	1.08	2.98	84.57	100.00
Spectrum 2	Yes	14.59		3.54	81.87	100.00
Spectrum 3	Yes	13.84	1.65	5.91	78.60	100.00
Spectrum 4	Yes	10.22		4.41	85.37	100.00
Spectrum 5	Yes	8.65		3.06	88.29	100.00
Spectrum 6	Yes	9.46		2.12	88.42	100.00
Spectrum 7	Yes	10.34		2.41	87.25	100.00
Spectrum 8	Yes	5.71			94.29	100.00

All results in weight [%]

The cleavage in separations morphology of DWTT fracture of 5# sample which preserved 4h under 600°C is shown in Fig. **7**. Massive initial cracks were detected on the direction which the crack travels. Inside these small initial cracks were the cementite precipitation particles. It is obvious that, the stress concentrated around the cementite precipitation particles and emerge initial cracks, more than several cracks began to grow and therefore caused the cleavage in separations and worsened the Low Temperature DWTT Toughness.

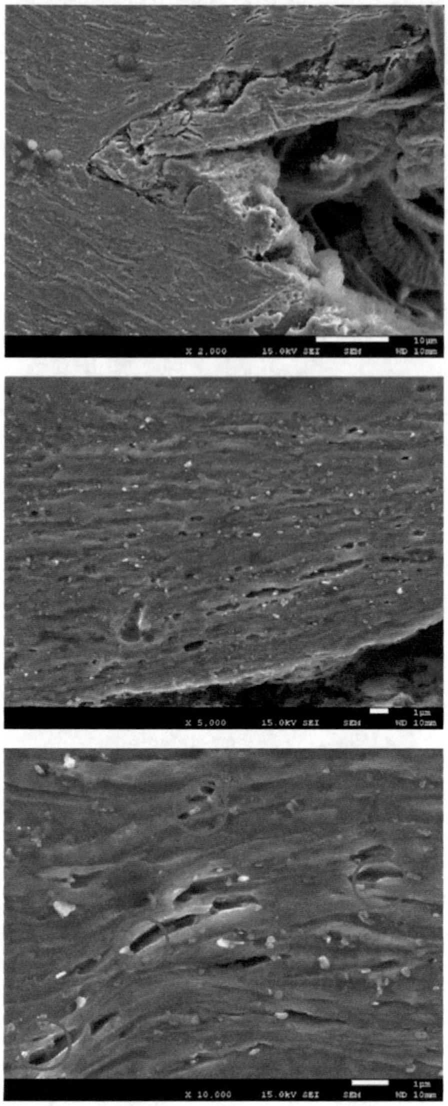

Fig. 7. Morphology of the cleavage in separations from DWTT fracture.

4 Summary

1) With the annealing temperature increases, the yield strength and the tensile strength of X70 pipeline steel both presented a trend from rise to decline. When the temperature reached 550 °C, the yield ratio experienced a remarkable increase.

2) With the annealing temperature increases, the -20 °C Impact Absorbed Energy decreases but the general level maintained in a fine condition. The -10 °C DWTT performance worsen when it is above 550 °C.

https://doi.org/10.1051/matecconf/201818602001

3）There is no significant change in microstructure when the annealing temperature is under 500 °C. When it is above 550 °C, massive cementite precipitation particles appear and the grain size grows up obviously which worsen the DWTT performance and caused the cleavage separations.

Reference

1. R Zheng. Shougang Science and Technology, **3**, 7~11(2003)
2. K Tong, C Zhuang, L Zhu etc. Materials Review, **24**, 98~101(2010)
3. N Narimani, B Zarei, H Pouraliakbar etc. Measurement, **62**, 97~107(2015)
4. L Li, M Zheng, L Sun etc. Welded Pipe and Tube, **33**,18~21(2010)
5. S Zhang, Y Ren, S Wang etc. *Effect of Temper Process for the Structure and Property of High Strength Pipeline Steel.* (Beijing: Metallurgical Industry Press, 2009)
6. W Zhao, L Zhao, Y Zhao etc. Materials for Mechanical Engineering, **39**, 28~31(2015)
7. Y Li, G Lin, L Ji etc. Materials for Mechanical Engineering, **34**, 14~18(2010)
8. G Li, H Xu, H Peng. Wide and Heavy Plate, **12**, 5~8(2006)
9. R Zhang, Y Liu. Anhui Metallurgy, **2** ,21~28(2009)

Effect of Shear Deformation on Texture Evolution in Rolled Pure Titanium

Huanping Yang and *Yaomian* Wang

School of Metallurgical Engineering, Xi'an University of Architecture and Technology, Xi'an 710055, China

Abstract. The cold rolling texture evolution under different strain states caused by coupled effects from different combinations of roll gap geometry and friction in pure titanium with initial typical recrystallized texture has been studied using viscoplastic self-consistent simulations. Under plane strain state, the texture is dominated by a typical cold rolling fiber texture RD//$\langle 10\bar{1}0\rangle$ in pure titanium. However, when shear deformation is induced and increased, RD//$\langle 10\bar{1}0\rangle$ fiber texture is decreased, whereas basal texture (0001)$[10\bar{1}0]$ is increased and a strong partial fiber texture RD//$\langle 2\bar{1}\bar{1}0\rangle$ is formed. The variation of cold rolling texture at different strain states can be ascribed to variation of relative contribution from activation of prismatic, basal and pyramidal slip.

1 Introduction

Pure titanium is widely used in many key fields such as aerospace, chemical, energy and ocean industries due to its high specific strength and excellent corrosion resistance [1]. However, because of the inherent low symmetry of the hexagonal close packed (HCP) structure, pronounced mechanical anisotropy is occurred in pure titanium during plastic deformation, which will significantly influence subsequent further forming of these metals into different shapes/parts [2]. The mechanical anisotropy can be decreased by controlling texture through plastic deformation [3]. As known, the deformed texture can be largely determined by strain state. And extensive researches have been conducted to reveal the relationship between strain state and deformation texture in FCC and BCC metals and showed that deformation texture was strongly influenced by shear deformation [4-6]. As for HCP metals, few researches have also reported that inducing shear deformation can change deformation texture [7-8]. However, studies on relationship between shear deformation and texture evolution in pure titanium are rarely reported.

Therefore, in the present work, the effect of the strain state focused on shear deformation caused by coupled effects of roll gap geometry and friction on rolling texture evolution in pure titanium was investigated using crystal plasticity simulation.

2 Simulation methods and parameters

The rolling texture under different strain states was simulated by means of viscoplastic self-consistent(VPSC) code. The detailed VPSC model and the predominant twin reorientation scheme are described in the work of Tomé [9] and Lebensohn et al.[10]. The Voce law, shown by Equation 1, is used to describe hardening behavior on different slip and twin systems.

$$\tau^s(\Gamma) = \tau_0 + (\tau_1 + \theta_1\Gamma)\left[1 - exp\left(\frac{-\theta_0\Gamma}{\tau_1}\right)\right] \quad (1)$$

$$\Gamma = \int_0^t \sum_s \left|\dot{\gamma}^s\right| dt \quad (2)$$

where Γ expressed in Equation 2 being the accumulated shear within time t on all slip systems, and $\left|\dot{\gamma}^s\right|$ is the shear rate on slip system s. In the simulation, the prismatic <a> slip $\{10\bar{1}0\}\langle\bar{1}2\bar{1}0\rangle$, the basal <a> slip $\{0001\}\langle\bar{1}2\bar{1}0\rangle$, the first-order pyramidal <c+a> slip $\{10\bar{1}1\}\langle\bar{1}\bar{1}23\rangle$ and two twinning modes, namely the tensile twinning $\{10\bar{1}2\}\langle\bar{1}011\rangle$ and compressive twinning $\{11\bar{2}2\}\langle\bar{1}\bar{1}23\rangle$, were taken into account. τ_0, τ_1, θ_0 and θ_1 are hardening parameters for each deformation system. The hardening behaviors represented by four parameters of the five deformation modes that lead to the best texture predictions are illustrated in Figure 1.

The different strain states were obtained by changing combinations of roll gap geometry and friction. As for rolling deformation, it is customary to assume plane strain state ε_{11}=-ε_{33} and all other components ε_{22}, ε_{12}, ε_{13}, ε_{23} are imposed to be zero. However, strong deviations from the plane strain prevail due to geometrical changes within the roll gap and the friction between roll surface and sheet surface, which leads to nonzero displacement gradient component e_{13} and e_{31}, resulting nonzero component ε_{13}. The different strain states are described by a ratio $\varepsilon_{13}/\varepsilon_{11}$ in the work. The shear deformation from geometrical and frictional effects is simplified as the material goes into and out the roll gap and is shown in Figure 2(a) and Figure 2(b), respectively. The shear deformation caused by roll gap geometry and friction is denoted as γ_g and γ_f, which is described by the ratio of shear e_{13} to e_{11} and e_{31} to e_{11}, respectively. According to Lee et al. [5], when shear deformation is caused just by roll gap geometry or friction, texture is nearly identical when $\gamma_g = \gamma_f$. And the geometrical effects and frictional effects can cancel each other. Therefore, the combined effects of roll gap geometry and friction are considered, and the deformation gradient is described by Equation 3.

$$E_{in} = \begin{vmatrix} 0.5 & 0 & -\gamma_g \\ 0 & 0 & 0 \\ \gamma_f & 0 & -0.5 \end{vmatrix} e_{11} \qquad E_{out} = \begin{vmatrix} 0.5 & 0 & -\gamma_g \\ 0 & 0 & 0 \\ \gamma_f & 0 & -0.5 \end{vmatrix} e_{11} \tag{3}$$

Four different state states from coupled effects of roll gap geometry and friction are listed in Table 1, i.e., $\varepsilon_{13}/\varepsilon_{11}$=0, $\varepsilon_{13}/\varepsilon_{11}$=0.3, $\varepsilon_{13}/\varepsilon_{11}$=0.5 and $\varepsilon_{13}/\varepsilon_{11}$=1. A true strain of 0.07 per pass has been imparted for all the samples to a total true strain of 0.7, corresponding to rolling reduction of 50%.

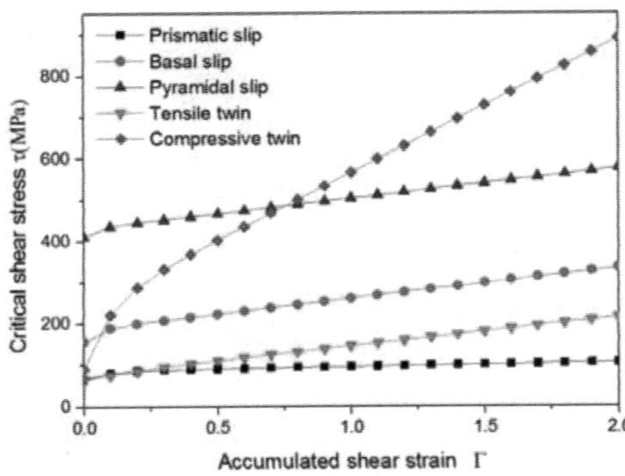

Fig. 1. Hardening behaviors of the slip/twin systems.

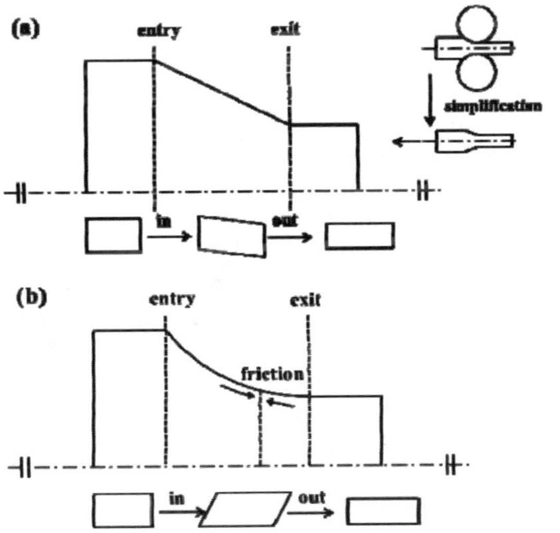

Fig. 2. Schematic diagram showing shear strain induced by roll gap geometry(a) and friction(b).

The texture evolution was characterized by the orientation distribution function (ODF) in the Euler space of (φ_1, ϕ, φ_2) following Bunge convention. When describing a specific orientation, the relationship between crystal and sample coordinate system is defined as RD=$[10\bar{1}0]$, TD=$[\bar{1}2\bar{1}0]$ and ND=[0001], respectively. The texture was calculated using MTEX toolbox [11] by imposing the orthorhombic sample symmetry for rolling process.

Table 1. Different strain states from different combinations of roll gap geometry and friction.

$\varepsilon_{13}/\varepsilon_{11}$	γ_g	γ_f
0	1	1
0.3	0.7	1
0.5	0.5	1
1	0	1

3 Results and Discussion

The initial texture is shown in Figure 3(a), which is a typical texture commonly observed in recrystallized pure titanium [12] and can be characterized by orientation range extended between the orientation B={0001}<$10\bar{1}0$>±40° and E={0001}<$11\bar{2}0$>±40° with peak intensity at component E. When the pure titanium is cold rolled by 50% at plane strain state, i.e., $\varepsilon_{11}=-\varepsilon_{33}$, $\gamma_g = \gamma_f = 0$, the texture is shown in Figure 3(b). In order to compare conveniently, Figure 3(c) gives the common orientations and fibers in cold rolled pure titanium. As shown, the main texture under plane strain state can be characterized by a typical cold rolling texture, i.e., RD//<$10\bar{1}0$> fiber texture, which verifies the reliability of the set of hardening parameters. And the RD//<$10\bar{1}0$> fiber texture is extending from component A (0001)$[10\bar{1}0]$ to component C ($\bar{1}2\bar{1}0$)$[10\bar{1}0]$ with maximum intensity near component B but shifting toward little ϕ at ($\bar{1}2\bar{1}5$)$[10\bar{1}0]$, which is a dominant texture commonly found in cold rolled pure titanium [13].

Figure 4 shows the texture in pure titanium cold rolled by 50% at different strain states. It can be seen that under coupled effects of roll gap geometry and friction and $\gamma_g=\gamma_f$, $\varepsilon_{13}/\varepsilon_{11}$ is equal to 0 because of just balanced effects, and the texture is similar to plane strain rolling texture shown in Figure 3(b). The results are in good agreement with that in face-centered cubic metals calculated by a fully constrained Taylor model, i.e., the simulated texture is identical to the plane strain simulation when $\gamma_g=\gamma_f$ [5]. When $\varepsilon_{13}/\varepsilon_{11}$ is increased to 0.3 due to unbalanced effects of roll gap geometry and friction, the texture characteristics have not changed much. When $\varepsilon_{13}/\varepsilon_{11}$ is further increased to 0.5, the texture is obviously changed. The dominant plane strain texture RD//<$10\bar{1}0$> fiber texture is significantly decreased, while the basal texture (0001)$[10\bar{1}0]$ is remarkably increased. The phenomenon that the basal texture is strengthened with increased shear deformation from VPSC simulation agrees well with the experimental results of Milner et al., who reported a ±30° transverse direction split basal texture was evolved to primarily basal texture when inducing shear deformation by isolated shear rolling [7]. And when $\varepsilon_{13}/\varepsilon_{11}$ is increased up to 1, the RD//<10$\bar{1}$0> fiber texture is further decreased and the basal fiber texture with maximum intensity at (0001)[10$\bar{1}$0] is further increased. Also, another partial fiber texture RD//<2$\bar{1}\bar{1}$0> with peak intensity close to component E but shifting to orientation (01$\bar{1}$3)[2$\bar{1}\bar{1}$0] is formed, which becomes the strongest component.

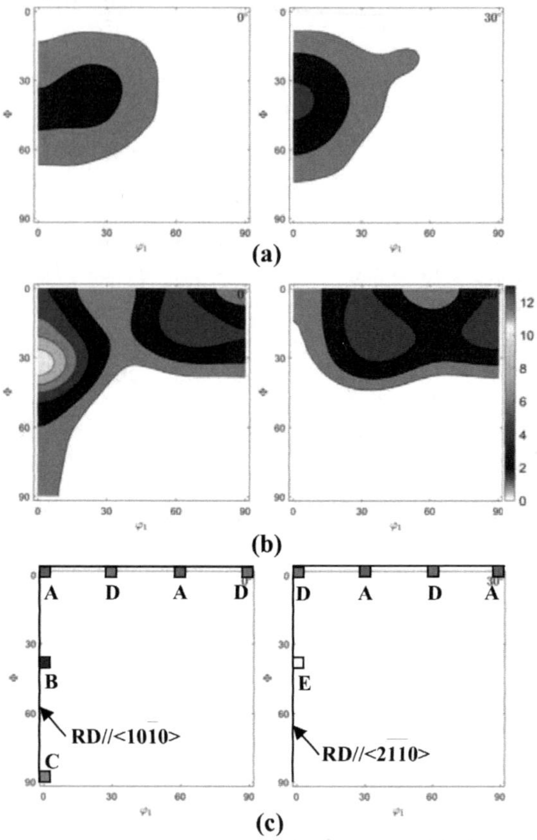

Fig. 3. Constant φ_2=0° and φ_2=30° of the ODF in initial pure titanium(a), cold rolled pure titanium under plane strain state(b) and common rolled texture components and fibers in pure titanium(c).

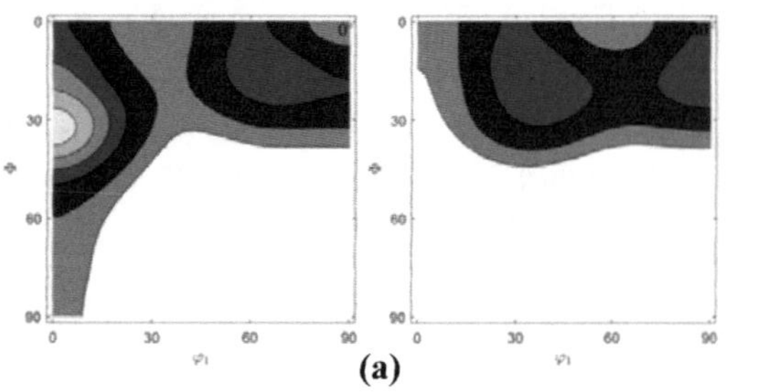

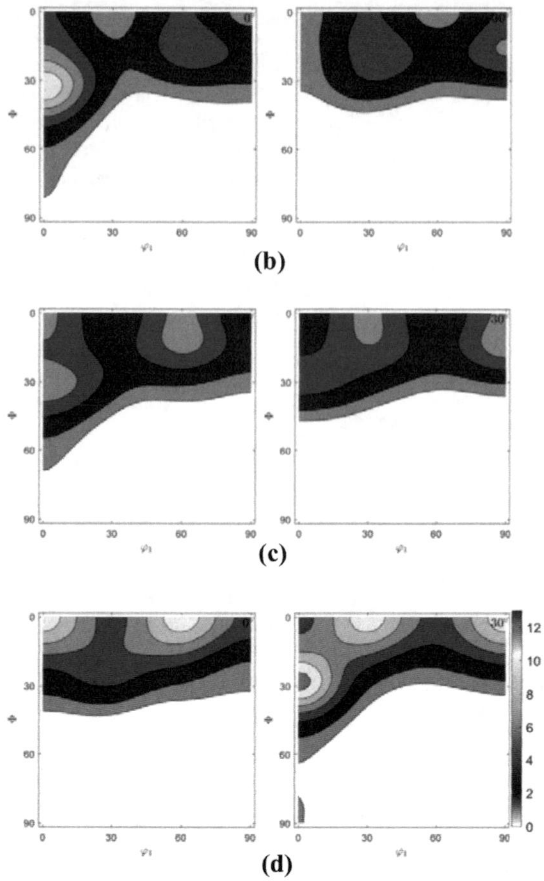

Fig. 4. Constant φ_2=0° and φ_2=30° of the ODF in cold rolled pure titanium at different strain states: (a) $\varepsilon_{13}/\varepsilon_{11}$=0, (b) $\varepsilon_{13}/\varepsilon_{11}$=0.3, (c) $\varepsilon_{13}/\varepsilon_{11}$=0.5 and (d) $\varepsilon_{13}/\varepsilon_{11}$=1.

The difference in cold rolled texture of pure titanium under different strain states is closely related to slip/twin activities. Figure 5 gives the slip/twin activities obtained from VPSC simulations. It shows that prismatic slip is the first dominant slip system under plane strain state, which contributes significantly to the formation of dominated RD//<10$\bar{1}$0> fiber texture. And basal slip is the secondary dominant slip system. Also, the activation fraction of the pyramidal slip and two twinning systems are relatively lower. When $\varepsilon_{13}/\varepsilon_{11}$ is increased to 0.3, the slip/twin activities have not changed much, which is consistent with the texture evolution shown in Figure 4(b). And when $\varepsilon_{13}/\varepsilon_{11}$ is increased to 0.5 and 1, the activation of slip/twin systems varies significantly. The activation of prismatic slip is decreased, whereas the activation of the basal and pyramidal slip is increased. The larger the $\varepsilon_{13}/\varepsilon_{11}$ is, the more the activation fraction of the prismatic slip, basal and pyramidal slip varies. The dominated near E component caused by pyramidal slip have also found in cold rolled submicron grain size pure titanium [14]. Certainly, because the texture evolution may be influenced by rolling reduction per pass, which is a true strain of 0.07 used in the paper, further researches should be conducted systematically.

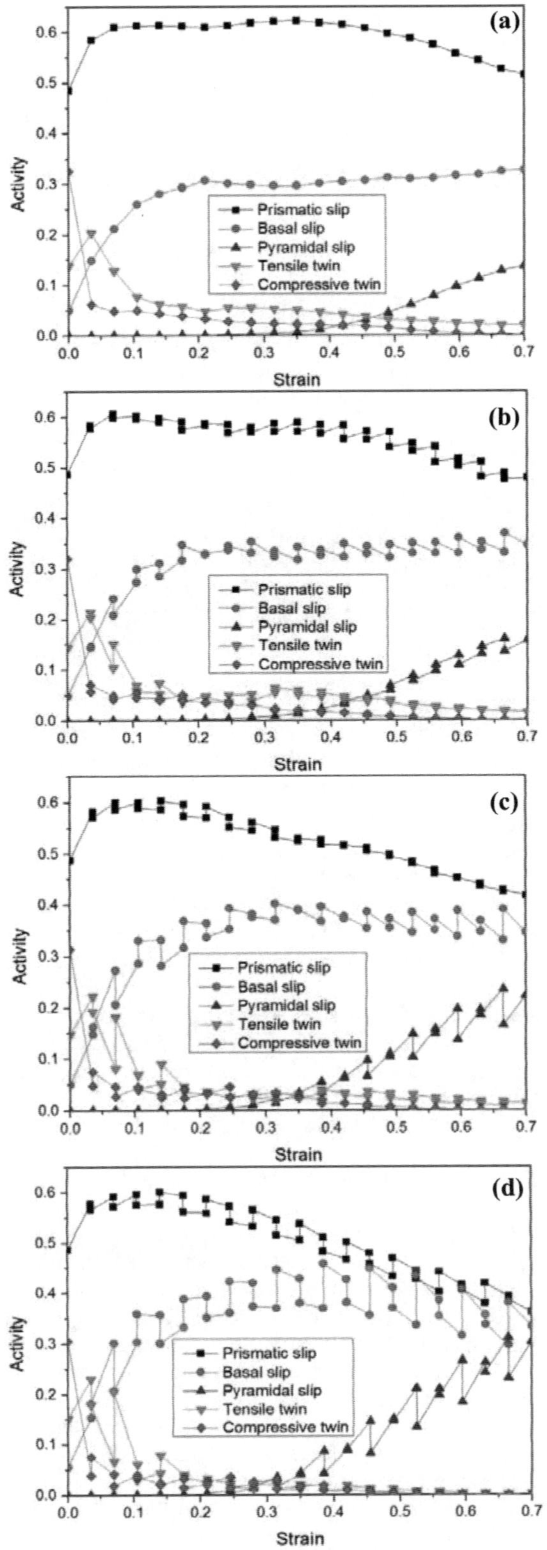

Fig. 5. Activity for slip/twin systems in cold rolled pure titanium at different strain states: (a) $\varepsilon_{13}/\varepsilon_{11}$=0, (b) $\varepsilon_{13}/\varepsilon_{11}$=0.3, (c) $\varepsilon_{13}/\varepsilon_{11}$=0.5 and (d) $\varepsilon_{13}/\varepsilon_{11}$=1.

4 Conclusions

(1) Strain state significantly influences cold rolling texture evolution in pure titanium. Under plane strain state, the texture is dominated by a typical fiber texture RD//$\langle 10\bar{1}0\rangle$ with maximum intensity at $(\bar{1}2\bar{1}5)[10\bar{1}0]$. When shear deformation is induced by coupled effects of roll gap geometry and friction and as the shear deformation is increased, the RD//$\langle 10\bar{1}0\rangle$ fiber texture is decreased, whereas the basal texture $(0001)[10\bar{1}0]$ is increased and a strong partial fiber texture RD//$\langle 21\bar{1}0\rangle$ with peak intensity at $(01\bar{1}3)[2\bar{1}\bar{1}0]$ is formed.

(2) The variation of cold rolling texture with increased shear deformation can be attributed to decreased activation of prismatic slip and increased activation of basal and pyramidal slip compared with that under plane strain state.

References

1. G. Lütjering, J.C. Williams, *Titanium*, second ed., Springer-Verlag, Berlin, 2007
2. H. Inagaki, Z Metallkd. **83**, pp40-46, (1992)
3. Z.S. Zhu, R.Y. Liu, M.G. Yan, C.X. Cao, J. Mater. Sci. **32**, pp5163-5167, (1997)
4. O. Engler, C.N. Tomé, M.Y. Huh, Metall. Mater. Trans. A **31**, pp2299-2315, (2000)
5. C.S. Lee, B.J. Bunge, Metall. Trans. A **22**, pp2637-2643, (1991)
6. K.H. Kim, D.N. Lee, Acta Mater. **49**, pp2583-2595, (2001)
7. J.L. Milner, F. Abu-Farha, T. Kurfess, V.H. Hammond, Mater. Sci. Eng. A **619**, pp12-25, (2014)
8. W.J. Kim, S.J. Yoo, H.T. Jeong, D.M. Kim, B.H. Choe, J.B. Lee, Scripta Mater. **64**, pp49-52, (2011)
9. C.N. Tomé, R.A. Lebensohn, U.F. Kocks, Acta Metall. Mater. **39**, pp2667-2680, (1991)
10. R.A. Lebensohn, C.N. Tomé, Acta Metall. Mater. **41**, pp2611-2624, (1993)
11. F. Bachmann, R. Hielscher, H. Schaeben, Solid State Phenom. **160**, pp63-68, (2010)
12. N. Bozzolo, N. Dewobroto, T. Grosdidier, F. Wagner, Mater. Sci. Eng. A **397**, pp346-355, (2005)
13. H.P. Lee, C. Esling, H.J. Bunge, Texture. Microstruct. **7**, pp317-337, (1988)
14. N.P. Gurao, S. Suwas, J. Mater. Res. **26**, pp523-532, (2011)

Strength Development and Microstructure of Alkali-activated Slag-MgO in Air Curing Condition

Chao-Lung Hwang[1], *Duy-Hai* Vo[1,2], *Mitiku Damtie* Yehualaw[1] and *Vu-An* Tran[3]

[1] Department of Civil and Construction Engineering, National Taiwan University of Science and Technology, Taipei, Taiwan
[2] Department of Civil Engineering, University of Technology and Education-The University of Danang, Danang, Vietnam
[3] Department of Civil Engineering, College of Engineering Technology, Can Tho University, Can Tho, Vietnam

Abstract. The aim of this study is to analysis the effect of MgO on strength development and microstructure of alkali-activated slag (AAS) in air curing condition. Four mixtures of AAS are prepared using different MgO content (0%, 5%, 10%, and 15 % by weight of slag) at water to binder ratio of 0.4. The flow, compressive strength, scanning electron microscopy, and X-ray diffraction are tested under relevant standards. The addition of MgO significantly accelerated the hydration rate of AAS. AAS with adding MgO tended to increase the compressive strength and to reduce the flow. The higher adding MgO content was associated with higher hydrotalcite-like phase (Ht) formation which improved the microstrure of AAS in the air curing condition.

1 Introduction

Ordinary Portland cement (OPC) has been widely used for the construction due to the many benefits from good properties to the cost of it. Based on the popular raw materials, the cement production can be easily conducted at many countries in the world. However, there is an estimation of about 1510 kWh consumed to produce one ton of cement as well as this process released about 900 kg of CO_2 into the atmosphere and approximately contributed about 5% of the global anthropogenic CO_2 production [1], [2].

In order to reduce the impacts into the environment as well as ensure sustainable development, using the alternative materials like as fly ash, ground granulated blast furnace slag (GGBFS), rice husk ash, lime stone powder and more to blend with OPC for reducing the consumption of OPC has focused in many researches [3]-[5]. On the other hand, alkali-activated slag (AAS) which is produced with 100 % GGBFS with activators such as sodium hydroxide, sodium silicate or sodium carbonate were considered as an alternative binder in the construction. This has certainly many advantages such as reuse the by-product, reduce the CO_2 emissions from cement productions, have similar or better properties to Portland cement. However, these materials have quick setting time and high drying shrinkage or micro-cracks. Different researchers had tried to mitigate the drying shrinkage of AAS. Bakharev et al. [6] used gypsum with 6 % by weight to reduce both of autogenous and drying shrinkage effects of AAS due to the ettringite (Aft) which is formed by the expansion of high voluminous of it. On the other hand, dolomite was also used as an expansion agent due to its content MgO, C_2S and a little amount of CaO. Xu Lingling et al. [7] showed that, the burning condition of dolomite and the dosage of MgO-based expansive increased the paste expansion. Expansion of cement increases with the lower burning temperature and the coarser particles size of MgO. Additionally, the expansion increased with increasing the burning condition but the expansion time delayed.

Using magnesia (MgO) to improve the properties of AAS has been studied for long times ago. In their study, Ben Haha et at. [8] showed the performance of AAS with various of MgO content in different slag. The results illustrated that the main hydration products are still C-S-H gel and hydrotalcite-like phase (Ht) which are formed from the reaction between MgO and slag. The formation of Ht increases with an increasing content of MgO. On the other hand, the higher compressive strength in the mixture with high MgO content. In another research, Fejin et al. [9] used two kind of reactive MgO to modify the alkali-activated slag and the results showed that the high reactive MgO produced Ht quickly and filled in to the macro-pores to form the meso-pores, hence it improves the compressive strength of AAS. Moreover, the high reactive MgO released more the Ht lead to the decreasing drying shrinkage. Comparing with the high reactive MgO, the medium reactive MgO also form the Ht and C-S-H in the hydration products. However, the rate of reaction of medium reactive MgO is slower than the high reactive MgO. Moreover, the reactive MgO were used as an activator for AAS to replace for conventional activators. Another research of Fei Jin et al. [10] used two kinds of reactive MgO various from 2.5% to 20% by weight to activate the GGBFS. The results gave hydrotalcite-like phase and C-S-H gel were also the main hydration products and the more reactive MgO released more hydration products which contribute to higher strength of AAS paste. The hydration products content increased with increasing the MgO content and curing time.

Even though, numerous researches have been done so far, still the problems are not fully resolved. This study investigated the effect of MgO on the fresh properties, strength development and microstructure of alkai-activated slag-MgO (AASM) in air curing condition. MgO was used with 5 %, 10 % and 15 % by weight of GGBFS, sodium hydroxide and sodium silicate with modulus SiO_2/Na_2O equal to 0.4 were used as an activator for AAS. The flow and compressive strength up to 28 days were used to estimate the mechanical properties of AASM as well as the XRD and SEM results were used to analysis the microstructure of AASM.

2 Materials and test method

2.1 Materials

Table 1. Characteristics of binder materials.

Items		S4000	MgO
Physical properties	Specific gravity	2.98	2.91
	Mean particle size	14.563	10.57
	Specific surface area (m^2/g)	1.44	1.62
Chemical composition (wt.%)	SiO_2	33.39	9.64
	Al_2O_3	14.39	-
	Fe_2O_3	0.19	0.15
	CaO	41.08	0.41
	MgO	7.22	89.2
	SO_3	0.11	0.75
	TiO_2	0.50	-
	K_2O	0.60	-

In this study, the GGBFS and MgO were collected from the companies located in Taiwan used to produce the AASM. Physical and chemical properties of these materials were examined and shown in Table 1. The mainly component of GGBFS was composed CaO (59.75%) and SiO_2 (26.74%), while MgO was pure material with 89.2% by weight. About the physical properties of these materials, MgO is presented a finer material than GGBFS based the lower mean particle size (10.57 µm compared with 14.563 µm, respectively) and higher specific surface area (1.62 (m^2/g) compared with 1.44 (m^2/g), respectively).

According to XRD patterns results in Figure 1, GGBFS is mainly found at amorphous phases, while MgO is pure and well crystallized material. On the other hand, MgO particle size was significantly smaller than GGBFS. In order to activate the AASM, a combination of sodium hydroxide (NaOH) with a purity higher than 98 % by weight and sodium silicate (Na_2SiO_3 with Na_2O: 8.26 %; SiO2: 25.7 % and water: 66.04 %) was used as an activator solution. Sodium hydroxide solution was mixed in water with molarity of 10M to dissolve the pellets before casting the samples. Local tap water was used in this study.

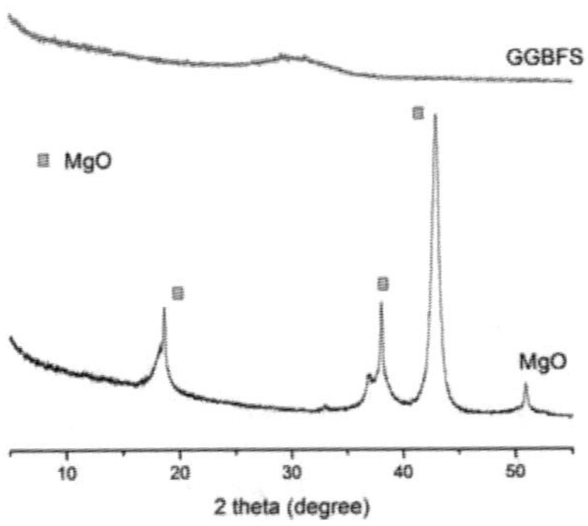

Fig. 1. XRD pattern of raw materials.

2.2 Mix proportions and test methods

The alkali-activated slag paste was prepared with proportions of adding MgO content of 0, 5, 10 and 15 % by weight of slag. All the parameters were kept constant, including the modulus SiO_2/Na_2O ratio of 0.4; the proportion of Na_2O of 4 % by the weight of binder, the water to solid equal to 0.4 and all of them were clearly shown in Table 2. First, sodium hydroxide and sodium silicate were mixed for two minutes to get homogenous activator solution before casting. Then, MgO was mixed with a part of water to homogenous for two 2 minutes and later slag and all the water were added. After 3 minutes, the activator solution was added and the AAS pastes resulted. For the fresh AAS paste, the flow test was conducted and 50x50x50 mm cubic paste samples were prepared with two layers as well as the mold was shaken by hand to limit the moisture content. To prevent the evaporation of water from the paste, all the samples were immediately covered by thin film and after 1 day, the samples were demolded and put into the chamber of 60% ± 1% humidity at 27 ± 2°C.

The compressive strength of AAS paste was measured on 1, 7, 28 days according to ASTM C 109 Microstructure analysis test was done for 1 & 28 age days specimens by using SEM. Finally, XRD test was conducted with a scanning range from 10 to 50 degree (2 theta) for 28 days of curing times.

Table 2. Mixtures design for the alkali-activated slag-Mgo.

Samples	Ingredient (%)					Flow (cm)
	w/b	GGBFS	MgO	SiO_2/Na_2O	Na_2O	
M0	0.4	100	0	0.4	4	34
M5	0.4	95	5	0.4	4	31.5

M10	0.4	90	10	0.4	4	29.5
M15	0.4	85	15	0.4	4	28

3 Results and discussion

3.1 Flow and compressive strength development

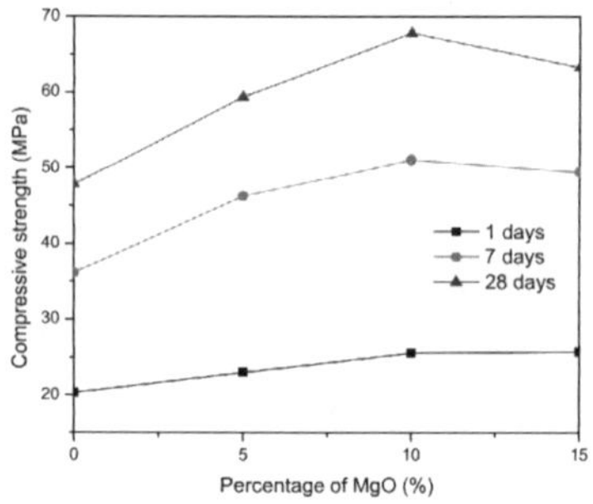

Fig. 2. Strength development of AASM.

The flow of AAS with various MgO contents is illustrated in Table 2. The flow reduced with increasing contents of the MgO. The flow of samples with 0%, 5%, 10%, and 15% MgO content were 34 cm, 31.5 cm, 29.5 cm, and 28 cm, respectively. The compressive strength of AASM pastes is shown in Figure 2. There is a slight increase in compressive strength of AASM for all proportions containing MgO on 1 day after demoulding. As expected, the paste with 15% MgO exhibits the highest compressive strength in 1 day. This result showed that high MgO content accelerated the hydration rate of AAS in the early age days [10]. At 28 days of air curing condition, the AASM showed the highest compressive strength with 10% MgO content. The compressive strength values of M5, M10, M15 were increased about 24.5 %, 42.5 %, 32.4 %, respectively, compared with mixture without adding MgO.

MgO accelerated the hydration reaction of AAS leading to more the hydration products formed [9], based on that, the compressive strength of AASM is higher than AAS without MgO content. Furthermore, the hydrotalcite-like phase is one of hydration products which is high voluminous to fill into the voids of aggregate to increase the compressive strength. However, the compressive strength of AASM increased with increasing MgO content up to 10 % and slightly reduced with 15% MgO content due to the large proportion of slag replaced by MgO lead to reduce the aluminosilicate content.

3.2 X-Ray diffraction analysis

The X-ray diffraction of AASM at 28 days of curing are portrayed by Figure 3. The results expressed that C-S-H gel and hydrotalcite-like phase are the two main hydration products which are shown on the amorphous phase so the peak are too low [11]. It is found that the Ht peak was disclosed at $2\theta\sim11.7^{\circ}$, the intensity of Ht peak raise with an increasing in MgO content and the more unhydrated MgO were observed at $2\theta\sim42.9^{\circ}$ comparing with the mixture without adding MgO. The formation of hydrotalcite-like phase was exhibited by quick reaction between MgO with the broken Al-O and Si-O in the slag so the higher Ht was formed with increasing the MgO contents. Finally, the shaper peak of C-S-H gel was found at $2\theta\sim29.5^{\circ}$[12].

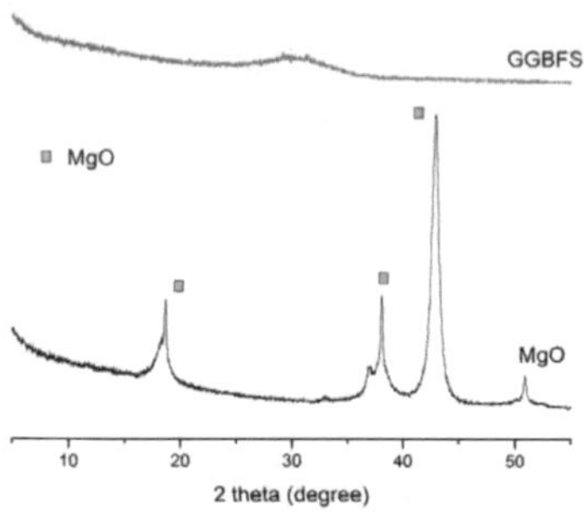

Fig. 3. XRD pattern of AASM at 28 age days.

3.3 SEM images

SEM images of AASM paste samples are presented by Figure 4 and 5. The results showed that after demoulding at 1 day, the AASM pastes exhibited high porous structures by un-hydrated slag and MgO particles. This result illustrated the low compressive strength in 1 day. Two main hydration products of AASM are C-S-H gel and hydratalcite which covered around the slag particles. The extent of formation of hydration products depends on the content of MgO and it increases with a higher proportion of MgO. Based on that the porosities of AASM paste samples lightly decreased. Especially, hydrotalcite-like phase was found with the mixtures contained adding MgO after demoulding and the amount of Ht increased with more the adding MgO contents lead to the higher compressive strength of mixtures contained more adding MgO.

Fig. 4. SEM image of AASM paste samples at 1 day (1) M0; (2) M5; (3) M10; (4) M15.

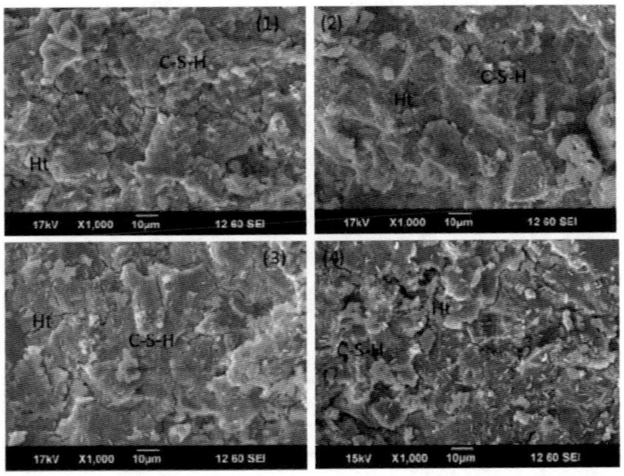

Fig. 5. SEM images of AASM paste samples at 28 days (1) M0; (2) M5; (3) M10; (4) M15.

After 28 days of curing, the denser structures were reported from all AASM samples due to consisting primarily of C-S-H gels. Fibrous Ht were detected in all AAS paste samples with and without adding MgO content which agreed with the XRD results. On the other hand, all the samples showed the small cracks in the structure because of high shrinkage of AAS.

4 Conclusions

Effect of various MgO contents into the alkali-activated slag on the fresh property, strength development and microstructure were studied. In this research, the AAS were added by 5%, 10% and 15% MgO by weight of slag to do the paste samples. The samples were tested the flow, compressive strength and microstructure based on XRD and SEM analysis test. Based on the results of investigation, the following conclusions are drawn:

(1) The MgO has little effect to accelerate hydration reaction of AAS in the early age and reduce in the flow of AAS paste.

(2) The optimum dosage of MgO which significantly increased the compressive strength of AAS in the air curing condition is 10% by weight of slag which contributed by hydrotalcite-like phase with high voluminous to fill into the macropores to form the mesopores hence the micro-structure becomes more denser.

(3) The XRD and SEM analyses illustrated the hydration products are mainly C-S-H gel and Ht-like phase to contributes the strength development of AASM. In addition, the microstructure of AASM becomes denser with increasing the curing times by growing up the crystallization of hydration products. On the other hand, increasing the adding MgO presents the dense structure by more hydration products. However, it is found to decrease the strength with increasing of MgO content (15% by weight) since reduce the aluminosilicate content as well as the more late Ht formation lead to form cracks matrix.

Using MgO to modify the mechanical properties and microstructure of alkali-activated slag potentially offer for reducing the impacts into environment as well as ensure the sustainable development. However, later Ht formation should be studied to limit the cracks by high voluminous of Ht. It is hoped that the more researches in this major will support for new greener materials.

Acknowledgment

The authors gratefully acknowledge the Hwang's research group at the National Taiwan University of Science and Technology (NTUST) for assistance in conducting experimental works

References

1. E. Worrell, L. Price, N. Martin, C. Hendriks, L. Ozawa-Meida, "Carbon Dioxide Emission from the Global Cement Industry", Annual Review of Energy and the Environment, vol. 26, pp. 303-329 (2001)
2. E. Kwon, J. Ahn, B. Cho, D. Park, "A study on development of recycled cement made from waste cementitious powder", Construction and Building Materials, vol. 83(Supplement C), pp. 174-180 (2015)
3. H. Chao-Lung, B.L. Anh-Tuan, C. Chun-Tsun, "Effect of rice husk ash on the strength and durability characteristics of concrete", Construction and Building Materials, vol. 25(9), pp. 3768-3772 (2011)
4. S. Kumar, R. Kumar, A. Bandopadhyay, T.C. Alex, B. Ravi Kumar, S.K. Das, S.P. Mehrotra, "Mechanical activation of granulated blast furnace slag and its effect on the properties and structure of portland slag cement", Cement and Concrete Composites, vol. 30(8), pp. 679-685 (2008)
5. M.M. Hossain, M.R. Karim, M. Hasan, M.K. Hossain, M.F.M. Zain, "Durability of mortar and concrete made up of pozzolans as a partial replacement of cement: A review", Construction and Building Materials, vol. 116, pp. 128-140 (2016)
6. T. Bakharev, J.G. Sanjayan, Y.B. Cheng, "Effect of admixtures on properties of alkali-activated slag concrete", Cement and Concrete Research, vol. 30(9), pp. 1367-1374 (2000)
7. X. Lingling, D. Min, "Dolomite used as raw material to produce MgO-based expansive agent", Cement and Concrete Research, 35(8), pp. 1480-1485 (2005)
8. M.B. Haha, B. Lothenbach, G. Le Saout, F. Winnefeld, "Influence of slag chemistry on the hydration of alkali-activated blast-furnace slag — Part I: Effect of MgO", Cement and Concrete Research, vol. 41(9), pp. 955-963 (2011)
9. F. Jin, K. Gu, A. Al-Tabbaa, "Strength and drying shrinkage of reactive MgO modified alkali-activated slag paste", Construction and Building Materials, vol. 51(Supplement C), pp. 395-404 (2014)

10. F. Jin, K. Gu, A. Al-Tabbaa, "Strength and hydration properties of reactive MgO-activated ground granulated blastfurnace slag paste", Cement and Concrete Composites, vol. 57(Supplement C), pp. 8-16 (2015)
11. H.A. Abdel-Gawwad, "Effect of reactive magnesium oxide on properties of alkali activated slag geopolymer cement pastes", Ceramics-Silikaty, vol. 59 (1), pp. 37-47 (2015)
12. S. Bernal, J. Provis, V. Rose, R. Mejia, "High-Resolution X-ray Diffraction and Fluorescence Microscopy Characterization of Alkali-Activated Slag-Metakaolin Binders", Journal of the American Ceramic Society, vol. 96 (6), pp. 1951-1957 (2013).

Comparison Study of Dynamic Elastic Moduli of Cement Mortar and No-cement Slag Based Cementitious Mortar Activated with Calcined Dolomite with Impulse Excitation Technique

Herry Suryadi Djayaprabha[1,3], *Ta-Peng* Chang[1] , *Jeng-Ywan* Shih[2]

[1]Department of Civil and Construction Engineering, National Taiwan University of Science and Technology (NTUST) (Taiwan Tech), Taipei 106, Taiwan, ROC.
[2]Department of Chemical Engineering, Ming Chi University of Technology, New Taipei City 243, Taiwan, ROC.
[3]Department of Civil Engineering, Parahyangan Catholic University, Bandung 40141, Indonesia.

Abstract. This paper presents the comparison of an experimental investigation on compressive strength and dynamic elastic moduli of mortars made of Ordinary Portland Cement (OPC) and ground granulated blast furnace slag (GGBFS) incorporating with calcined dolomite. Dolomite powder calcined at temperature 900°C emerged as a GGBFS activator for producing cementitious mortar binder. In this study, no-cement mortar is made by activating GGBFS with calcined dolomite by a fixed amount of 20 wt%. The compressive strengths and dynamic elastic moduli were measured at 7 and 28 days. Comparing with cement mortar, the compressive strength of no-cement mortar was found about 54.4 and 46.9% lower at ages of 7 and 28 days, respectively. Non-destructive evaluation of the dynamic elastic moduli was investigated by impulse excitation technique (IET). It measures the resonant frequencies of induced vibration signal in the flexural and torsional mode for determining the dynamic Young's modulus and the dynamic shear modulus. The Poisson's ratio was calculated by the dynamic Young's modulus and the dynamic shear modulus relationship. The results showed that the 28-day dynamic Young's and shear moduli of cement mortar were 31.91 and 14.43 GPa, respectively. The dynamic Young's and shear moduli of no-cement mortar were lower by 23.3 and 15.2% than that of cement mortar at the age of 28 days. The obtained results showed that the 28-day Poisson's ratio of no-cement mortar had a wider range between 0.177 and 0.209 than that of cement mortar which ranged from 0.180 to 0.185.

1 Introduction

Ordinary Portland cement (OPC), as a vital construction material, has reached worldwide annual production of 2.8 billion tonnes and expected to increase 4 billion tonnes per year [1]. Meanwhile, approximately 5-7% of global carbon dioxide (CO_2) emission originate from OPC production, ranging from 0.66 to 0.82 kg of CO_2 emitted for every kilogram produced [2]. On the contrary, a large amount of industrial waste by-products such as ground granulated blast furnace slag (GGBFS) is being generated every year throughout steel industries. It is reported that the annual worldwide production amount of GGBFS is 530 million ton every year [3]. With the similar chemical and mineral composition to OPC, GGBFS can be potentially used as cement replacement. The usage of GGBFS as cement replacement not only reducing the CO_2 emission but also solving industrial by-product waste disposal problem.

GGBFS without an activator does react slowly with water, so it may take longer time for a pure slag concrete made with GGBFS as the binder to reach the equivalent the 28-day compressive strength of concrete made with OPC [4]. The concrete and mortar binder made with 100% GGBFS plus activator, known as alkali-activated slag, was investigated intensively during the past few decades [5-7]. Chemical activators are commonly used as slag activators such as sodium hydroxide (NaOH), sodium silicate (Na_2SiO_3), and sodium carbonate (Na_2CO_3) [7].

Activation of GGBFS without chemical activator is feasible. Recent studies [8], [9] showed that dolomite powder ($CaMg(CO_3)_2$) being calcined at a temperature of 800-1000°C potentially used as GGBFS activator. The two-step decomposition of dolomite [10] shown as follows:

$$CaMg(CO_3)_2 \rightarrow CaCO_3 + MgO + CO_2 \quad (1)$$

$$CaCO_3 \leftrightarrow CaO + CO_2 \quad (2)$$

The formation of calcium carbonate ($CaCO_3$) and magnesium oxide (MgO) (Eq. 1) occurred at a lower temperature (Eq. 1), whereas the formation calcium oxide (CaO) (Eq. 2) takes place at a higher temperature. Several studies [11]-[13] showed that both CaO and MgO were used as potential activators for producing a GGBFS-based no-cement binder.

The impulse excitation technique (IET) is widely used techniques for determining the vibration behavior of the engineering material by measuring the resonant frequencies by non-destructive testing [14]. The IET consists in vibrating a specimen by exiting it with an impulse, capture the vibration signal with a microphone, and calculate the dynamic elastic moduli from the measured lowest flexural and torsional frequencies [15]. Some studies have shown the application of IET for determining dynamic elastic properties on building materials, for instance, the concrete at high temperatures [16], engineered stone [14], and granites [17].

The incorporation GGBFS and calcined dolomite to produce no-cement mortar still lack sufficient studies. Therefore, the objective of this study was to investigate the compressive strength and dynamic elastic moduli of no-cement mortar produced by the GGBFS activated by calcined dolomite (slag-dolomite) at the amount of 20 wt% (percentage by weight) and compare the results to cement mortar.

2 Experimental program

2.1 Materials

In the present study, type I Ordinary Portland Cement (OPC) produced by Taiwan Cement Co. Ltd., ground granulated blast furnace slag (GGBFS) obtained from CHC Resources Corporation, Taiwan, and dolomite powder, obtained from Taimax Material Co. Ltd., Taiwan, were used. Calcined dolomite was obtained by laboratory calcination, it was calcined at complete decarbonation temperature 900°C for 60 minutes [9]. The physical properties and chemical composition of OPC, GGBFS, and dolomite powder are given in Table 1. The aggregate used to produced mortar was natural river sand (<4.75 mm) with the fineness modulus (FM) of 2.99 and saturated surface dry specific gravity of 2.72.

2.2 Mix proportions and fresh mortar properties

Our previous study [9] has shown that the addition of 20 wt% calcined dolomite as GGBFS activator induced the highest compressive strength. In this study, the calcined dolomite amount was fixed by 20 wt% for producing no-cement slag-dolomite mortar. The mortar produced consist of 1 part binder and 2.75 parts of sand proportioned by mass according to ASTM C109 [18]. Mix proportions were designed by a volumetric method with a fixed water-to-binder ratio of 0.4 presented in Table 2.

The workability of mortar was attained by using the optimum dosage of superplasticizer (Type F polycarboxylic ether). Flow table test was conducted according to ASTM C1437 [19] to determine the optimum dosage of superplasticizer to obtain a mortar flow of 110±5% in 25 drops. The fresh properties with a satisfy workability of slag-dolomite mortar is shown in Figure 1.

Table 1. Physical properties and chemical composition of OPC, GGBFS, and raw dolomite.

Parameters	**OPC**	**GGBFS**	**Dolomite**
SiO_2, %	20.42	33.54	2.71
Al_2O_3, %	4.95	13.00	-
CaO, %	61.96	39.85	90.36
K_2O, %	-	-	0.23
MgO, %	3.29	6.90	1.84
SO_3, %	2.40	1.67	-
Fe_2O_3, %	3.09	0.50	0.52
L.O.I, %	1.75	4.54	4.34
Specific Gravity	3.15	2.83	2.70

Table 2. Mortar mixtures (units: kg/m^3).

Code†	**Water**	**OPC**	**GGBFS**	**CD**	**Sand**	**SP**
M-OPC	230	575	-	-	1552	2.86
M-SD	226	-	453	113	1558	0.52

Notes: M-OPC = ordinary Portland cement mortar, M-SD = slag-dolomite mortar, CD = calcined dolomite, SP = superplasticizer (type F)

2.3 Mortar specimens preparation

The 50-mm cubic and 100×50×25 mm platy mortar specimens were made in triplicate to be tested for determining compressive strength and dynamic elastic moduli. Mortar specimens were cast by pouring fresh mortar mixture into the cubic molds, which then was tamped by hand to ensure the complete consolidation of the mortar, and towel finished. All samples were demolded after 24 hours and cured in saturated lime water according to ASTM C511 [20] at a temperature of 25±2°C.

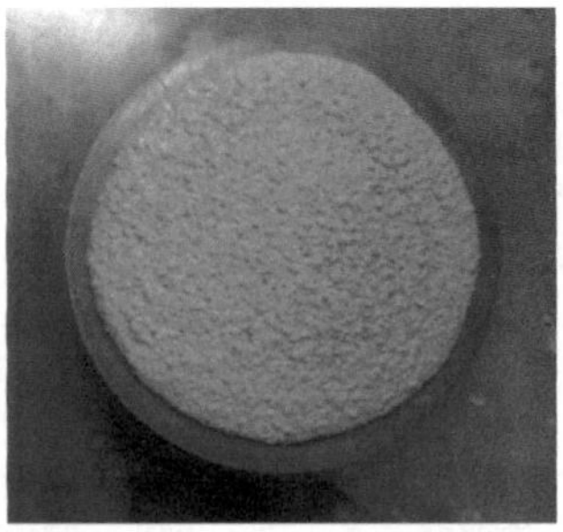

Fig. 1. Workability of the slag-dolomite mortar.

2.4 Testing method

The compressive strength test of 50-mm cubic slag-dolomite mortar specimens was determined by uniaxial compression according to ASTM C109 [18] using a computer-controlled hydraulic compression testing machine with a 2000-kN capacity.

The RFDA Basic instrument (Resonant Frequency and Damping Analyzer, developed by Integrated Material Control Engineering, Belgium) was used to analyze the dynamic elastic moduli of slag-dolomite mortar specimens by the impulse excitation method (IET) according to ASTM E1876 [21]. The IET measures the resonant frequencies in order to calculate the dynamic elastic moduli on the non-destructive basis.

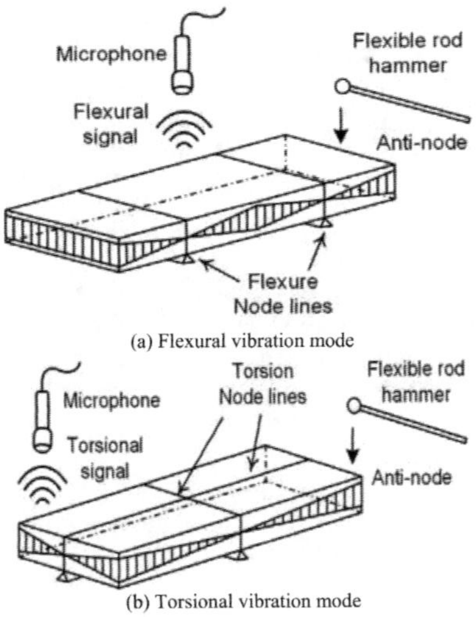

Fig. 2. Schematic representation of both flexural and torsional vibration modes applied in platy specimen [16].

Li et al. [22] confirmed that the dynamic elastic moduli of OPC mortar platy specimens with aspect ratio 2:1 give an error about 0.11-1.64% comparing with the impact resonance method. In this study, the 100×50×25 mm platy specimen with aspect ratio 2:1 was positioned on the testing support to create flexure (Figure 2a) and torsion (Figure 2b) node lines. The specimens were gently tapped on the anti-node by a flexible rod hammer by 0.7-mm sphere ball in order to induce both flexural and torsional vibration modes. The vibration signal was then captured by a 16 kHz microphone and sent to RDFA software. The software extracted the data by using Fast Fourier Transformation (FFT) in order to determine both flexural (f_f) and torsional (f_t) frequencies. The dynamic Young's modulus (E_{dyn}) and shear modulus (G_{dyn}) was calculated by using Eq. 3 and Eq. 4 respectively.

$$E_{dyn} = 0.9465\left(\frac{m \cdot f_f}{w}\right)\left(\frac{l^3}{t^3}\right)T \qquad (3)$$

$$G_{dyn} = \left(\frac{4 \cdot l \cdot m \cdot f_t^2}{w \cdot t}\right)\left(\frac{B}{(1+A)}\right) \qquad (4)$$

$$\nu_{dyn} = \frac{E_{dyn}}{2G_{dyn}} - 1 \qquad (5)$$

where m, l, w, and t represents specimen mass (g), length, width, and thickness (mm), f_f and f_t are the flexural and torsional frequency (Hz), and T, A, and B are correction coefficients which can be found in ASTM E1876 [21]. Dynamic Poisson's ratio (ν_{dyn}) was determined from the relationship between dynamic Young's modulus and shear modulus as shown in Eq. 5.

3 Results and discussion

3.1 Comparison of compressive strength

The compressive strength tests of hardened cement mortar (M-OPC) and slag-dolomite mortar (M-SD) were carried out at ages of 7 and 28 days. The compressive strength comparisons are graphically represented in Figure 3. The compressive strength of slag-dolomite mortar at ages of 7 and 28 days reached about 24.4 and 30.6 MPa, which was lower about 54.4 and 46.9% lower than cement mortar. The compressive strength of slag-dolomite mortar proved to be increased with curing time due to GGBFS hydration process. The positive effect from GGBFS activated by calcined dolomite showed that after 7 days of hydration the compressive strength of slag-dolomite mortar was approximately 79.6% of that at the age of 28 days. The enormous difference of compressive strength between cement mortar and slag-dolomite mortar related to the distinctive hydration products.

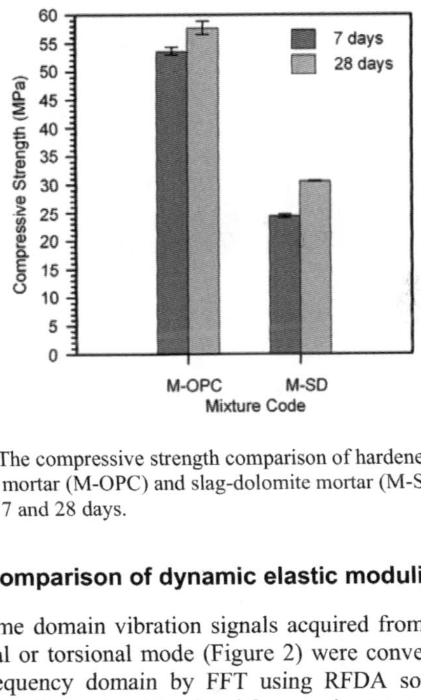

Fig. 3. The compressive strength comparison of hardened cement mortar (M-OPC) and slag-dolomite mortar (M-SD) at ages of 7 and 28 days.

3.1. Comparison of dynamic elastic moduli

The time domain vibration signals acquired from either flexural or torsional mode (Figure 2) were converted to the frequency domain by FFT using RFDA software. Then, the flexural and torsional frequencies were used to calculate dynamic Young's (E_{dyn}) and shear modulus (G_{dyn}). The dynamic Poisson's ratio (ν_{dyn}) was directly calculated from the dynamic Young's modulus and the dynamic shear modulus relationship as shown in Eq. (5). The comparison of dynamic elastic moduli and Poisson's ratio of cement mortar and slag-dolomite mortar are shown in Figure 4. The dynamic Young's modulus was 32.2 and 33.9 GPa, while the shear modulus was 13.8

and 14.3 GPa, for cement mortar at ages of 7 and 28 days, respectively. For slag-dolomite mortar, the dynamic Young's modulus was lower by 25.8 and 14.4% while the shear modulus was lower by 23.3 and 15.2% at ages of 7 and 28 days, respectively. As a porous material, the 28-day Poisson's ratio of cement mortar was found in the range of 0.180 to 0.185, whereas the 28-day Poisson's ratio of slag-dolomite mortar had a wider range between 0.176 and 0.209.

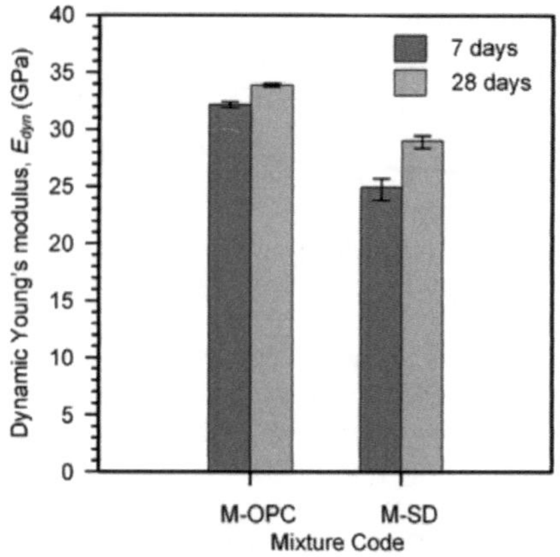

(a) Dynamic Young's modulus

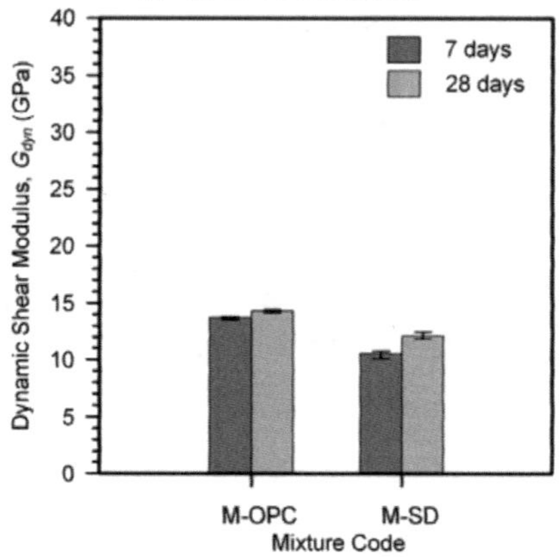

(b) Dynamic shear modulus

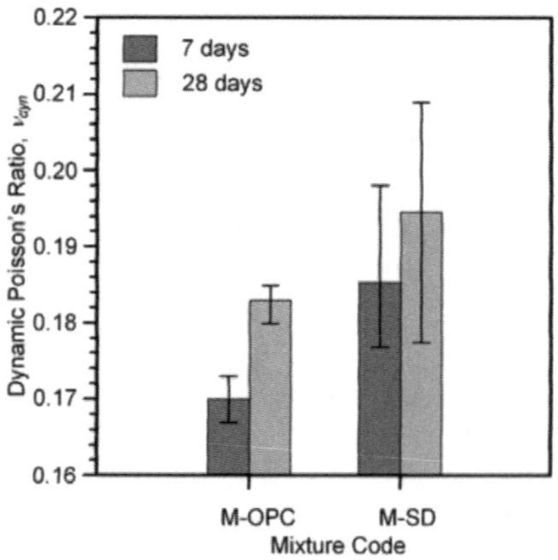

(c) Dynamic Poisson's ratio

Fig. 4. The dynamic elastic moduli and Poisson's ratio comparison of hardened cement mortar (M-OPC) and slag-dolomite mortar (M-SD) at ages of 7 and 28 days.

4 Conclusion

The present study evaluated the comparison of compressive strength and dynamic elastic moduli of hardened cement mortar and no-cement slag-dolomite mortar. The following conclusions were drawn:

(1) The compressive strength of cement mortar was 53.5 and 57.7 MPa at ages of 7 and 28 days, respectively, while slag-dolomite mortar reached 24.4 and 30.6 MPa at ages of 7 and 28 days.

(2) Associated with the compressive strength results, it can be observed that no-cement mortar made with GGBFS incorporating with calcined dolomite had lower dynamic elastic moduli than cement mortar. The dynamic Young's modulus of slag-dolomite mortar was lower by 25.8 and 14.4% while the shear modulus was lower by 23.3 and 15.2% at ages of 7 and 28 days.

(3) Mortar composed of a binder and fine aggregate that lead to a porous material. The Poisson's ratio of porous material varied based on its porosity. From results obtained, it can be observed that the slag-dolomite mortar had a wider range of Poisson's ratio (0.177-0.209, at age of 28 days) than that of cement mortar (0.180-0.185, at age of 28 days). Therefore, it can be concluded that the cement mortar had less porosity than slag-dolomite mortar.

Acknowledgement

The authors would like highly acknowledge the financial aids from both the Ministry of Science and Technology through the grants of MOST 106-2221-E-011-056 and the National Taiwan University of Science and Technology (Taiwan Tech) to conduct this study.

References

1. M. Schneider, M. Romer, M. Tschudin, H. Bolio, Cem. Concr. Res., **41**, pp. 642-650, (2011).
2. L. K. Turner and F. G. Collins, Constr. Build. Mater., **43**, pp. 125-130, (2013).
3. H. Zhao, W. Sun, X. Wu, B. Gao, J. Clean. Prod., **95**, pp. 66-74, (2015).
4. S. Mindess, J. F. Young, D. Darwin, *Concrete, 2nd Edition*, Pearson Education Taiwan Ltd., pp. 103, (2008).
5. F. Puertas, B. Gonzalez-Fonteboa, I. Gonzalez-Taboada, M. M. Alonso, M. Torres-Carrasco, G. Rojo, F. Martínez-Abella, Cem. Concr. Comp., **85**, pp. 22-31, (2017).
6. C. Atiş, CahitBilim, ÖzlemÇelik, OkanKarahan, Constr. Build. Mater., **23**, pp. 548-555, (2009).
7. S.-D. Wang, K. L.Scrivener, P.L.Pratt, Cem. Concr. Res., **246**, pp. 1033-1043, (1994).
8. K. Gu, F. Jin, A. Al-Tabbaa, B. Shi, Constr. Build. Mater., **68**, pp. 252-258, (2014).

 https://doi.org/10.1051/matecconf/201818602004

9. H. S. Djayaprabha, T.-P. Chang, J.-Y. Shih, C.-T. Chen, Constr. Build. Mater., **150**, pp. 345–354, (2017).
10. R. Otsuka, Thermochim. Acta., **100**, pp. 69-80, (1986).
11. Y. Jeong, H. Park, Y. Jun, J. H. Jeong, J. E. Oh, Cem. Concr. Comp., **72**, pp. 155-167, (2016).
12. M. S. Kim, Y. Jun, C. Lee, J. E. Oh, Cem. Concr. Res., **54**, pp. 208-214, (2013).
13. F. Jin, K. Gu, A. Al-Tabbaa, Cem. Concr. Comp., **57**, pp. 8-16, (2015).
14. J. P. L. d. Santos, P. M. Amaral, A. C. Diogo, L. G. Rosa, Key Eng. Mater., **548**, pp. 220.-230, (2013).
15. J. I. Etcheverry, G. A. Sánchez, J. Sound Vib., **321**, pp. 631-646, (2009).
16. O. Bahr, P. Schaumann, B. Bollen, J. Bracke, Mater. Design, **45**, pp. 421-429, (2013).
17. V. Pires, L. G. Rosa, A. Dionísio, Constr. Build. Mater., **64**, pp. 440-450, (2014).
18. ASTM C109/109M-16a, *Standard Test Method for Compressive Strength of Hydraulic Cement Mortars (Using 2-in. or [50-mm] Cube Specimens)*, **04.01**, (2016).
19. ASTM C1437-15, *Standard Test Method for Flow of Hydraulic Cement Mortar*, **04.01**, ASTM C1437-15, (2015).
20. ASTM C511-09, *Standard Specification for Mixing Rooms, Moist Cabinets, Moist Rooms, and Water Storage Tanks Used in the Testing of Hydraulic Cements and Concretes*, **04.02**, (2009).
21. ASTM E1876-15, *Standard Test Method for Dynamic Young's Modulus, Shear Modulus, and Poisson's Ratio by Impulse Excitation of Vibration*, **03.01**, (2015).
22. C.-L. Li, C.-W. Yang, T.-P. Chang, Presented at Taiwan Concrete Institute (TCI) 2017 Conference on Concrete Engineering, Chiayi, Taiwan, November 23-24, (2017) (In Chinese)

Damages Assessment of Cement-Paste and Cobblestone Interfacial Transition Zone by Acoustic Emission

Tam NGUYEN-TAT[1,2], *Trung-Hieu* TRAN[2], *Narintsoa* RANAIVOMANANA[1], *Jean-Paul* BALAYSSAC[1]

[1] LMDC, Université de Toulouse, INSA, UPS, France
[2] Faculty of Civil Engineering, Hanoi Architectural University, Vietnam

Abstract. In cementitious materials, bonding at the contact surface between cement-paste and coarse aggregates plays an important role and determines the durability of the material. Recent studies have evaluated the impact strength on this contact surface by experiments on different surface roughness aggregates and concluded that the contact area between the two components was the weakest zone in concrete. This paper presents experiment test on cement-paste and cobblestone contact surface to determine the types of damage modes using Acoustic Emission Technique. The result is that the shear stress occupies only about 10% of the total stress experienced on the contact surface.

1. Introduction

In concrete, the bond between the cement paste and the coarse aggregate is significant because it affects to the strength of the material. Study on the affections of aggregate surface on adhesion force with cement paste and mortar in the experiments under compress, tensile and shear samples have been published recently [1]-[5]. And damages at the contact area between cement paste and coarse aggregate or Interfacial Transition Zone (ITZ) are interested the authors in those experiments. By damage analysis of ITZ, some remarkable results have been introduced.

The failure mechanism of the ITZ between cobblestone aggregate and mortar was investigated by under the combined of normal and shear stresses [2]. Prism specimens were casted in 50×50×200 mm3 molds with marble aggregate placed in the molds and the interface inclination angles were 45 degrees. It was concluded that the shear bond strength of the interface was approximated to the shear strength of the mortar.

The affection of surface roughness to the bonding strength of the ITZ between coarse aggregates and mortar and on the damage processes of concrete under uniaxial tensile and uniaxial compression were presented [3]. Five kinds of surfaces have made: fully polished, medium polished, saw-cut, sandblasted, and notched surfaces. As conclusion, when the normal stress lower than 60% of the mortar compressive strength, the tensile and shear bond strengths of the interface increased with increased surface roughness of aggregates, and also resulted in rising of concrete splitting tension strength.

The shear failure process of the ITZ of the granite - mortar specimens have been monitored by the Acoustic Emission (AE) system [4]. Cylinder specimens with 64mm diameter including three layers: mortar, granite and mortar were conducted by direct shear tests. Failure processes at the contact surface are recorded with the AE signals obtained. The parameters of these signals showed that the AE technique was appropriate for the experiments and it also indicated the damage severity of the samples.

Recently, the mechanical properties of the cement paste - aggregate specimens were experimentally studied [5]. These samples were composed of limestone aggregates and Portland cement paste which dimension of 10×10×30 mm3. The results showed that the cement-aggregate interface is the weakest zone in concrete, this is good agreement with that in [6]. The authors also concluded that if the normal force is applied to the shear samples, the shear strength value increases significantly.

In general, the above ITZ studies between cement paste and aggregate have assessed the affections of surface roughness to bonding as well as on the tensile and compressive strengths of the material at the interfaces. Therefore, studies on the types of damage modes occurring on the surface of contact between the two materials need further study. In this paper, the author presents an experiment test to determine the damage modes on cement paste exposed to cobblestone by Acoustic Emission Technique (AET). The results showed that the stresses at ITZ include both modes I and mode II, and the proportion of mode I is significantly higher than that in mode II.

2. Experimental methodology and acquirement system

2.1 Specimen setup

https://doi.org/10.1051/matecconf/201818602005

The specimen namely CPRI (Cement-Paste and Rock Interface) has dimension of 60×60×120 mm3 and consists of two parts; the lower part is a block of cobblestone, a common aggregate for concrete, with a 45 degree surface that created by saw cut. The above cement-paste is poured into the lower part at the incline surface and maintained in the laboratory 28 days before conduct the experiment test. In cement-paste, the type I Portland cement [7] was used with the water/cement ratio is 3.0. The mechanical properties were determined at 28 days on three 118×225 mm2 cylinders. Compressive strength (f'_c) of 68.9 MPa was assessed by direct compression tests based on EN 12390-3 [8]; the Elastic modulus of 23.6 GPa was determined according to RILEM CPC8 recommendation [9].

The specimen was subjected to compression test to verify the fracture modes in the cement-paste and cobblestone interface under monotonic loading. To produce shear damage at cement-paste and cobblestone contact (shear damage under compression test), notches with 20mm depth were created surround this interface boundary. The loading was controlled by a rate 50 N/s.

2.2 Acoustic Emission acquirement setup

The Acoustic Emission (AE) is a common non-destructive testing method that used to (i) locate the micro-cracks, (ii) characterize the damage modes, and (iii) to assess the damage severity of the construction structures. Some AE parameters can be directly extracted from one signal such as the Count, the Amplitude, the Duration, the Rise-Time, the Energy, etc. These are called primary parameters [10]. By damage mode classification applying RA method, the mode I cracks result in waveforms with short Rise-Time and high Average Frequency whereas mode II cracks result in waveforms with longer Rise-Time and lower Average Frequency. The Rise-Time and the maximum Amplitude are used to calculate RA value, while the Average Frequency value is obtained from AE count and the Duration time [11].

AE activities were recorded by using eight-channel PCI-8 acquisition device of the Physical Acoustic Corporation (PAC). AE detection was carried out by R15- sensors PAC series with the specifications as indicated in Table 1. These sensors are mounted on the surface of the specimens with silicon grease as coupling agent, and they were placed close to the expected location of the future cracks path to minimize errors in the AE event localization.

Table 1. Specification of the R15-series.

Parameters	Operating specification
Operating frequency	50 – 400 kHz
Resonant frequency	150 kHz
Peak sensitivity	80 dB
Dimensions	19 mm (diameter) × 22.4 mm (height)

The PAC preamplifiers model 2/4/6 (gain selectable 20/40/60 dB + 5% dB) apply a gain of 40 dBs to eliminate the background noise. The acquisition system was calibrated before each test using a standard source Pencil Lead Break (PLB) procedure Hsu-Nielsen, and to verify that nothing has changed on sensors sensitivity before and after the test, the Auto Sensor Test was performed. In these tests, the AE events are located by applying the wave propagation velocity of 3,900 m/sec. AE detection was performed by means of six sensors that fixed on the sample as presented in (a) Testing specimen (b) Specimen at failure

Fig. 1(a). The three-dimension Cartesian coordinates of the sensors are indicated in Table 2. (a) Testing specimen (b) Specimen at failure

Fig. **1**(b) introduces the sample after the experiment; the slip damage was occurred at the contact surface between cement-paste and cobblestone and split the sample into two parts.

(a) Testing specimen (b) Specimen at failure

Fig. 1. Sensors arrangement on specimen (a), and specimen at failure (b).

Table 2. AE sensors arrangement.

No.	X (cm)	Y (cm)	Z (cm)
1	3	8	6
2	0	10	3
3	0	5	3
4	3	7	0
5	6	9	3
6	6	4	3

3. Result and discussion

3.1 Evolution of mechanical damage versus AE hits

In term of mechanical, the Fig. **2** indicates the evolution of loading versus axial displacement for the tested specimen. In the first segment, loading is linear developed that representation of the elastic behavior of material, then it becomes non-linear and reaches the maximum value at final of the test. The peak loading value is 14.3 kN corresponding to axial shortening of 177 m.

https://doi.org/10.1051/matecconf/201818602005

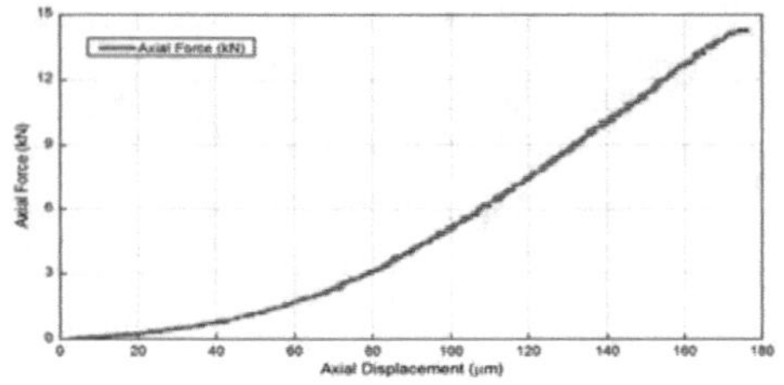

Fig. 2. Evolution of loading and displacement on CPRI specimen.

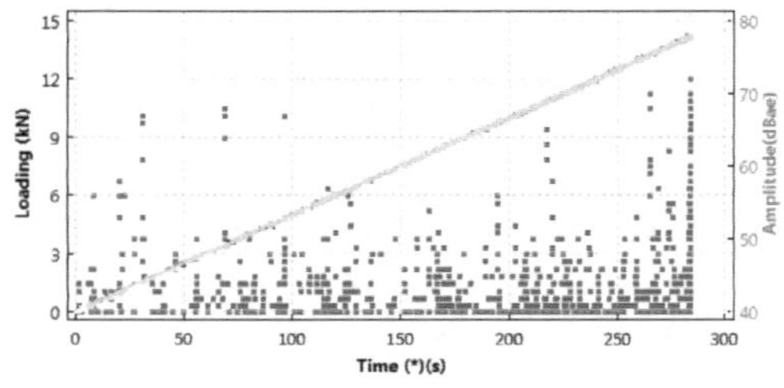

Fig. 3. Correlation of Loading and Amplitude versus time in CPRI test.

In term of AE, the Fig. **3** shows the loading curve and the number of signals obtained in the test versus time. Accordingly, AE signals were obtained immediately after the loading began but the AE number is varied during the experiment. It is noteworthy that at the peak load moment, although this is a very short time but the number of AE hits was recorded very significant and accounted for 282/984 or 28.7% of the total number of signals obtained during the experiment. The high abundant signals were collected at the peak load corresponds to the rapidly growing of the crack after the loading reaches its maximum value. In this test, specimen generated 984 signals after time t = 292 s.

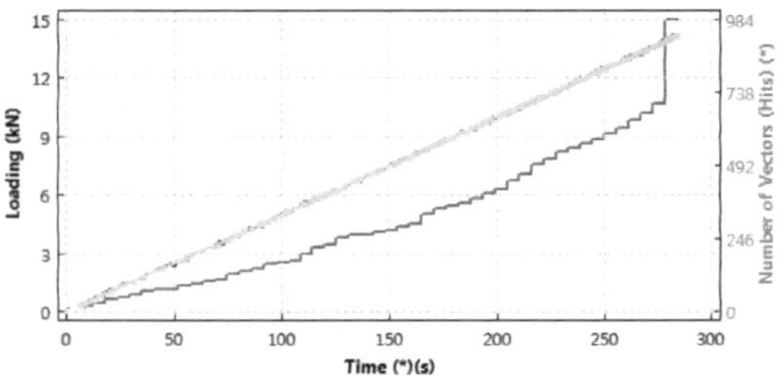

Fig. 4. Loading and cumulative AE hits vs. Time.

The amount of AE signals obtained during the test is also represented by a cumulative curve as shown in Fig. **4**. In the figure, the cumulative AE hit (the red line) is stable increased along with the development of the axial loading (the blue line). It increased dramatically at the end of the experiment when the load reached the max value and the sample was completely failure.

The number of AE signals received in the six sensors in Fig. **5** showing that the sensors CH1 to CH6 have recorded different number of signals. The CH6 obtained the most (363 signals) while CH4 obtained the least (70 signals). This difference may be due to the variation of the distance between the events and the sensors, because the sensors that farther from the events will receive lower number of signals than the closer sensors. In addition, this discrepancy may due to the influence of the notches on the sample surface to the transmission path and to reduce the amplitude of the signals leading to some sensors have not detect the signals.

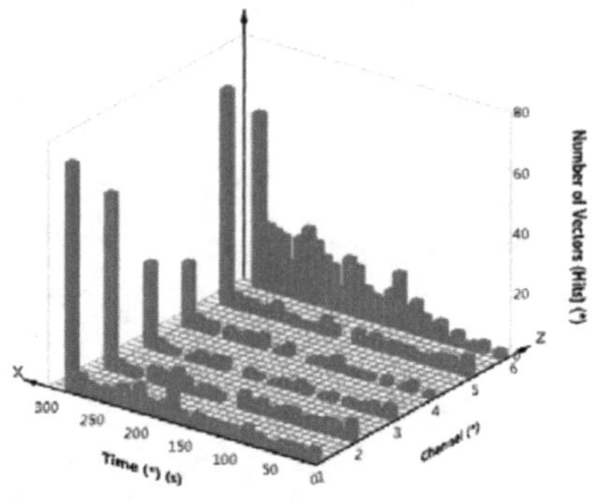

Fig. 5. Number of AE hits in Channel CH1 to CH6 vs. Time.

The cumulative ABEN (Absolute Energy) value of the signals obtained in this experiment is shown in Fig. **6**. It was found that the ABEN cumulative value from the beginning of the test to the before failure was about 150,000 aJ, and at the peak load moment (equivalent to 28.7% of the received signals), ABEN cumulative value increased to 335,991 aJ. Obviously, at the final stage before the sample is completely destroyed, the AE signals have high Amplitude as well as Energy, which may be due to large-scale fracture (or macro-crack) occurrence in the specimen.

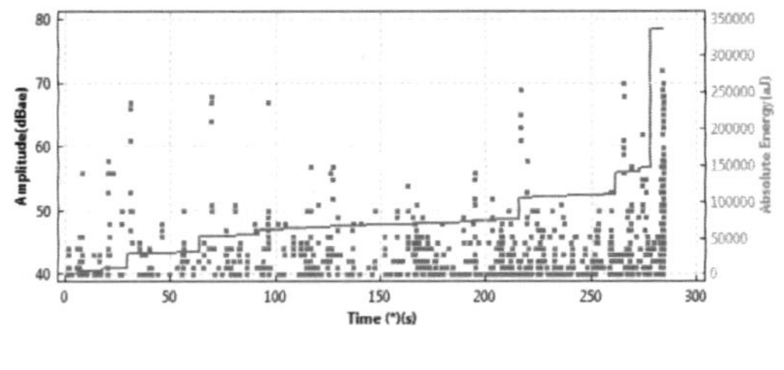

Fig. 6. Signal Amplitude and Cumulative ABEN vs. Time.

3.2 Evolution of RA and AF versus damage modes

In the experiment tests on the cementitious material applying AE technique, a damage modes classifying method using AE parameters such as Average Frequency (AF, kHz) and RA (Rise-Time per Amplitude, ms/V) is presented in RILEM TC 212-ACD [12]. In this CPRI test, the RA and AF values are calculated and displayed on the same graph as shown in Fig. **7**. Accordingly, the low RA and high AF signals are located in Zone 1, while the high RA and low AF signals are classified in Zone 2. Also in Fig. **7**, the signal in Zone 1 represents 71.7% of the total signal.

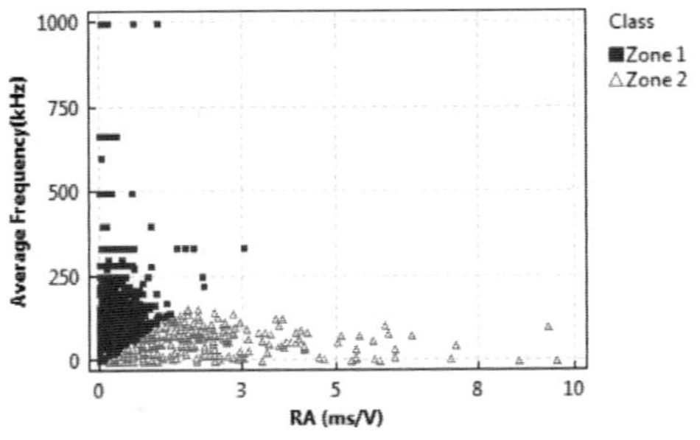

Fig. 7. Correlation between RA and AF.

To assess the other parameters of Zone 1 and Zone 2, the correlation diagrams between Rise-Time (RT) and AF, between Amplitude and AF are shown below. In Fig. **8**, the relationship between RT and AF introduces that signals are classified into two distinct areas as Zone 1 and Zone 2. This indicates that, instead of using RA and AF values, the RT and AF can be used to classify destructive modes. Signals with low RT and high AF are assigned to Zone 1 and the ones with large RT and small AF are assigned to Zone 2.

In term of Amplitude (Amp, dB), signals in Zone 1 have higher Amp value than those in Zone 2. In Fig. **9**, Zone 1 signals have Amp ranging from 40 to 80 dB while Zone 2 from 40 to 60 dB. For low Amp signals that classified into Zone 2, it is noted worthy that the distance from the sensor to the original source or event may influence to the Amp value. Because the sensors that farther from the event than others may receive low Amp signals due to amplitude attenuation versus travelling distance [13].

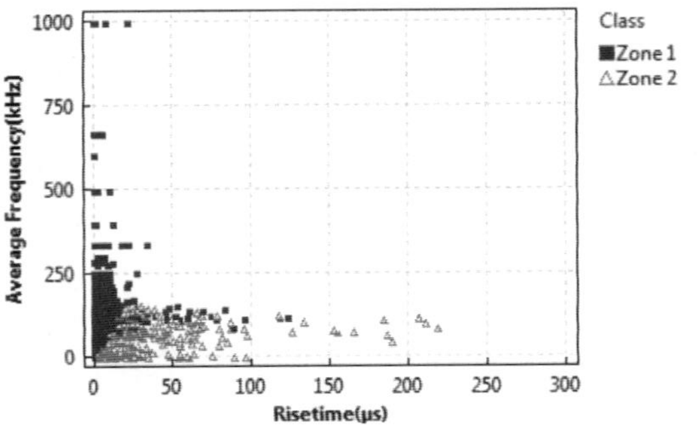

Fig. 8. Correlation between RT and AF.

The following content related to the occurrence of the signals in Zone 1 and Zone 2 versus the experimental time. The Fig. **10** shows that the signals from both groups were appeared from the beginning of the experiment, and the sudden accelerate at the moment before the final failure. In the graph, the RT value of Zone 2 is higher than that of Zone 1.

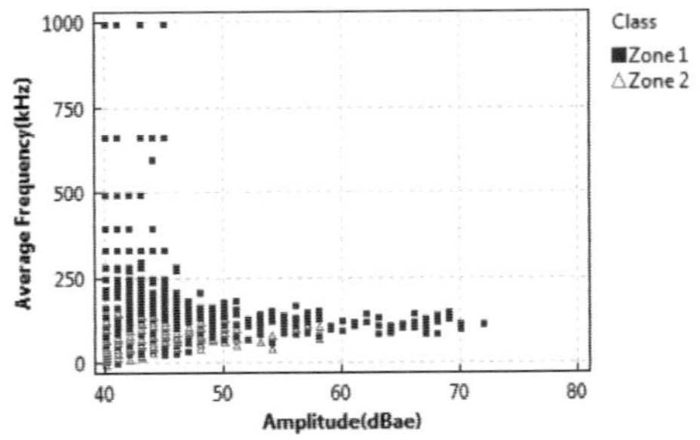

Fig. 9. Correlation between Amplitude and AF.

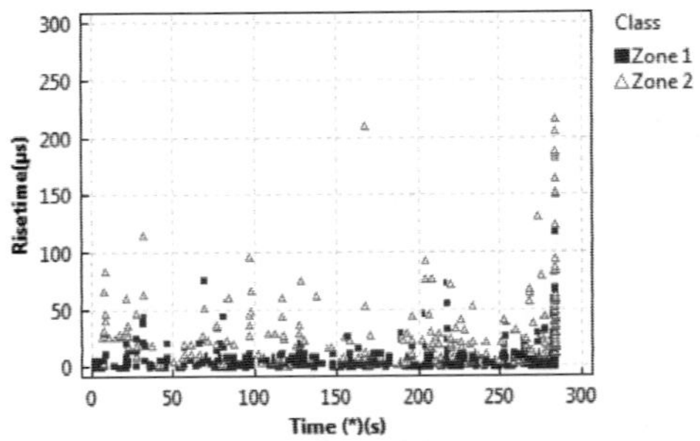

Fig. 10. Evolution of RT vs. time.

3.3 Evolution of moving average 100-hit RA and AF versus damage modes

In the light of RILEM TC 212-ACD [12], the shear-induced damage under compression test on CPRI were performed to verify the stress proportion at ITZ between cement-paste and cobblestone. The RA method is applied to classify the damage modes in the samples. And as recommended by [12], the RA and AF values should be calculated from moving average of more than 50 continuous hits. In this test, authors obtained these two features by determining the moving average of 100 continuous signals.

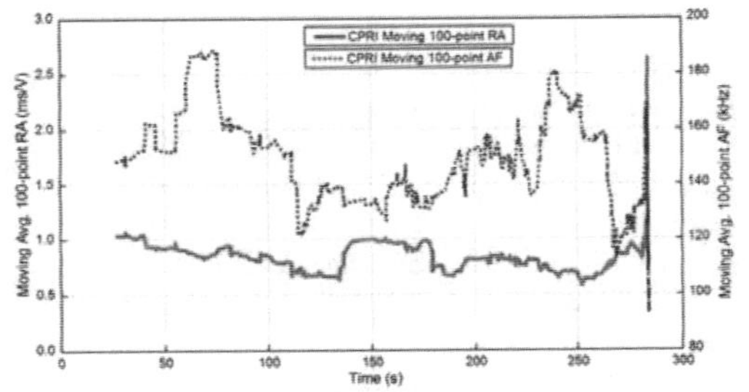

Fig. 11. Evolution of moving average continuous 100-point RA and AF.

The Fig. **11** shows the variation of RA and AF that determined from the moving average 100 continuous points in the CPRI test. As can be seen, the first phase, from the beginning to the moment before the final failure, the RA values are fluctuated in a range from 0.5 to 1.2 ms/V. At the final phase, the RA rises dramatically to 2.7 ms/V and the sample is completely failure. In term of

AF feature, in the first stage, AF fluctuated in the range of 120 to 190 kHz. At failure period, AF dropped rapidly to 90 kHz and the CPRI sample was completely failure. Thus, before the final failure, the RA value increased and higher than 1.2 ms/V while AF dropped and lower than 120 kHz corresponding to the failure moment in Fig. **11**. The damage mode in this final period corresponding to shear mode that good agreement with [14].

3.4 Crack classification using RA method: results and analysis

The testing process in the CPRI is divided into two phases in Fig. **12**; in Phase 1, RA fluctuates in the range of 0.5 to 1.2 ms/V and AF is in the range of 120 to 190 kHz. In the next Phase, RA increases from 1.2 ms/V to 2.7 ms/V, while AF decreases from 120 kHz to 90 kHz. At this Phase, the cracks grow rapidly and the sample completely failure. Thus, in the shear-induced test on CPRI specimen, the damages in the first Phase can be classified as mode I, while mode II is classified to final Phase. The breakdown of two damage phases is shown in Fig. **12**, where the diagonal line has a ratio K = 3/300 = 1/100. The intersection points at the two phases are the transition between the two damage states, from mode I to mode II or from mode II to mode I.

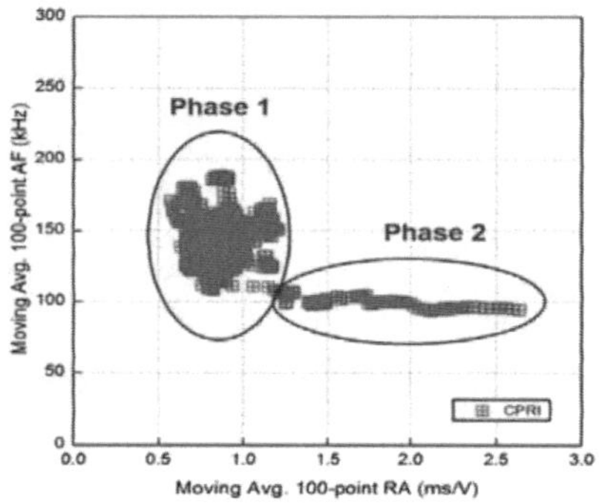

Fig. 12. Damage classification in tensile specimens.

If we accept the damage classification by the transition from mode I to mode II at the ratio of 1/100 as indicated by the diagonal line on the Fig. **12**, the following proportion of damage modes in tested specimen can be obtained. Phase 1, from beginning to before final state, the damage state is pure mode I. At failure moment (Phase 2), shear mode appears and quickly accelerate then dominant of 100% of the stresses. Comparing to the total received signals; the shear mode is occupied 10% of the total damage modes.

4. Conclusion

Aim to produce shear failure in cement-paste and cobblestone interfacial specimen under compression loading; the pre-notches are placed surround the contact surface to promote shear fracture. However, by the classification method based on RA and AF which determined from the moving average 100 moving signals, the results show that the cracks in the tested specimen are caused by the dominant tensile stress. The shear stress appeared at the final moment, in which the sample was completely failure, and the rate of mode II was about 10% to the total of the signals that obtained during the test.

References

1. W. A. Tasong, C. J. Lynsdale, J. C. Cripps, Aggregate-cement paste interface Part I. Influence of aggregate geochemistry, C.C.R. 29 (1999) 1019–1025.
2. X. Gu, L. Hong, Z. Wang, F. Lin, Experimental study and application of mechanical properties for the interface between cobblestone aggregate and mortar in concrete, C.B.M. 46 (2013) 156–166.
3. L. Hong, X. Gu, F. Lin, Influence of aggregate surface roughness on mechanical properties of interface and concrete, C.B.M. 65 (2014) 338–349
4. Y. Wang, S. J. Chen, H. T. Zhao, Y. Z. Chen, Acoustic emission characteristics of interface between aggregate and mortar under shear loading, Russ. J. Nondestruct. Test. 51 (2015) 497–508.
5. M. Jebli, F. Jamin, E. Malachanne, E. Garcia-Diaz, M.S. E. Youssoufi, Experimental characterization of mechanical properties of the cement-aggregate interface in concrete, C.B.M. 161 (2018) 16–25.
6. J. J. Zheng, C. Q. Li, X. Z. Zhou, Thickness of interfacial transition zone and cement content profiles around aggregates, Mag. Concr. Res. 57 (2005) 397–406.
7. ASTM C150/C150M – 15, Standard Specification for Portland Cement (2015)
8. European Standard, EN 12390-3, Testing hardened concrete - Part 3: Compressive strength of test specimens (2012)
9. RILEM TC CPC 8, Modulus of elasticity of concrete in compression, (1975)
10. C. Grosse, M. Ohtsu, eds., Acoustic Emission Testing, Springer Berlin Heidelberg, (2008)
11. RILEM Technical Committee, Test method for classification of active cracks in concrete structures by AE, Mater. Struct. 43 (2010) 1187–1189.
12. K. Ohno, M. Ohtsu, Crack classification in concrete based on acoustic emission, C.B.M. 24 (2010) 2339–2346.
13. D.G. Aggelis, A.C. Mpalaskas, D. Ntalakas, T.E. Matikas, Effect of wave distortion on acoustic emission characterization of cementitious materials, C.B.M. 35 (2012) 183–190.
14. A. Behnia, H. K. Chai, T. Shiotani, Advanced structural health monitoring of concrete structures with the aid of acoustic emission, C.B.M. 65

 https://doi.org/10.1051/matecconf/201818602006

Parameters Analysis for Bending Properties of Externally Prestressed Concrete Beam with Corrugated Steel Webs

Qiang Xu[1], *Jianyong* Song[1], *Jianxiang* Feng[2] and *Jinsheng* Du[2]

[1]Bridge technology research center, Research Institute of Highway Ministry of Transport, Beijing, China
[2]Department of Bridge engineering, Beijing Jiaotong University, Beijing, China

Abstract. Based on the bending moment-curvature method, a calculation program for the whole bending process of externally prestressed concrete beam with corrugated steel webs is compiled and parametric analysis is conducted. Considering the nonlinearity of material, this procedure can calculate the whole bending process of externally prestressed composite beam with corrugated steel webs under different load, cross section, and external tendons profiles and output the cross section strain of concrete and ordinary reinforcements, stress increment of external prestressing tendons and mid-span deflection. The results show that the tension steel area has the greatest impact on the bending performance of simply supported externally prestressed composite beam with corrugated steel webs; The second is concrete strength and effective external prestress; The compression steel area has limited impact on the bending performance of such beams. The results can provide a reference for design of the beam.

1 Introduction

In prestressed concrete (PC) bridges with corrugated steel webs, the steel webs mainly resist shear while the flanges mainly resist bending. Their advantages include lightness and high shear buckling strength of steel webs, efficient prestressing of concrete and easy maintenance. The appearance of Cognac Bridge of this type in France has prompted research and development efforts in various countries on this new composite structure [1]-[4]. Currently, many investigations have been done on such beam's ultimate capacities, stress distribution, failure mechanism, while few have addressed the factors that may influence the bending behavior of the beam [5]-[7]. In this paper, based on the moment-curvature method, a nonlinear full process program is established for externally PC beams with corrugated steel webs. By using this numerical program, parametric analysis is conducted to further investigate the factors that may affect the bending performance of the beam.

2 Calculation program

2.1 Calculation assumption

In the program, some assumptions as follow are adopted: Plane sections remain plane; The flexural rigidity of corrugated steel webs is negligible [8]-[10]; The corrugated steel webs with sufficient buckling strength, will not happen any form of buckling failure; With sufficient shearing strength, the beam only happens bending failure; Stress evenly distributed along externally prestress tendons; After cracking, the contribution of concrete in tension zone is negligible; The beam is incompressible vertically.

2.2 Material constitutive model

In the program, the following materials constitutions are adopted: The Rüsch equation is adopted to describe the unidirectional compression constitution of concrete, and the ultimate compressive strain of concrete is taken as 0.003. The ideal elastic-plastic model is used as the constitution of reinforcement steel, and the steel plate and the ultimate strain of the steel is taken as 0.01.The constitution of the externally prestressed tendons is given by:

$$\begin{cases} \sigma_p = E_p\varepsilon_p & \text{for } \varepsilon_p \le f_{py}/E_p \\ \sigma_p = f_{py} + 0.3E_p(\varepsilon_p - f_{py}/E_p) & \text{for } \varepsilon_p > f_{py}/E_p \end{cases} \quad .(1)$$

where, σ_p is the stress in prestressed tendons; E_p is the elastic modulus of prestress tendons; ε_p is the strain in prestress tendons; f_{py} is the yield strength of prestress tendons.

2.3 Cross-section balance equation

For any cross-section of the beam, we can always get the following equations:

$$\varepsilon_y = \begin{cases} \varphi(y + c - h) & \text{for } \varphi \ge 0 \\ \varphi(c - y) & \text{for } \varphi < 0 \end{cases} . \quad (2)$$

$$N_c + N_t + \sigma_s A_s + \sigma_g A_g + \sigma_s' A_s' + \sigma_p A_p = 0 \,. \quad (3)$$

$$\begin{aligned} &M + M_c + M_t + \sigma_s' A_s' (y_x - a_s') - \sigma_s A_s (y_x - a_s) \\ &+ \sigma_g' A_g' (y_s - a_g') + \sigma_g A_g (y_x - a_g) + \sigma_p A_p y_p = 0 \end{aligned} \,. \quad (4)$$

where, M is the total applied bending moment; N_c is the sum of concrete stress in compressive zone; M_c is the sum of bending moment of concrete stress in compressive zone about the centroidal axis of cross-sections; N_t is the sum of concrete stress in tensile zone; M_t is the sum of bending moment of concrete stress in tensile zone about the centroidal axis of cross-sections; The geometrical characteristic of cross-section is shown in Figure 1.

With (2), (3) and (4), if the stress increments of externally prestress tendons $\Delta\sigma_p$ are known, φ and c can be obtained.

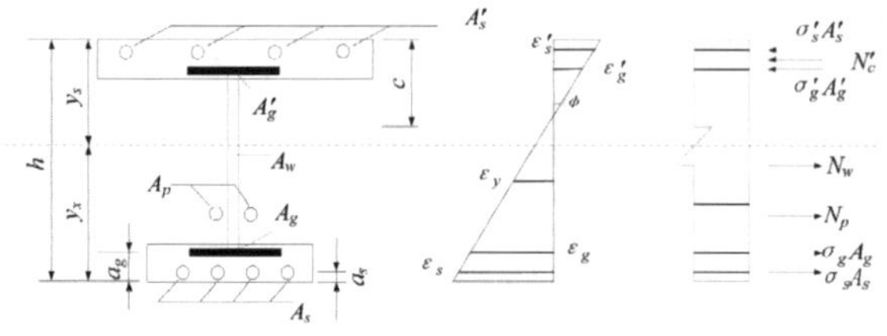

Fig. 1. Stress-strain relationship at a cross section.

2.4 Calculation of stress increments of externally prestressed tendons $\Delta\sigma_p$

The length increments of externally prestressed tendons ΔL_p have to respect compatibility with the deformation of the beam. Taking a beam of straight tendons with anchorages at the end diaphragms and one intermediate diaphragm as deviators for example, which is shown in Figure 2, ΔL_p can be expressed as:

$$\Delta L_p = \sqrt{f_c^2 + (\frac{1}{2}L + \mu_C - \theta_A \cdot y_p^A)^2} + \sqrt{f_c^2 + (\frac{1}{2}L + \mu_B + \theta_B \cdot y_p^B - \mu_C)^2} - L \quad (5)$$

and $\Delta\sigma_p$ can be expressed as:

$$\Delta\sigma_p = E_p \frac{\Delta L_p}{L_p} - \sigma_{pe} \quad (6)$$

where, f_c is the deflection of intermediate deviator location C; L is the span of the beam; θ_A , θ_B is the rotation angles at the anchorage locations A and B, respectively; μ_B , μ_C is the axial deformation at the anchorage location B and intermediate deviator location C, respectively; L_p is initial length of the prestressed tendons; σ_{pe} is the initial stress of the prestressed tendons; y_p^A and y_p^B is the distance from centroidal level to anchorage locations A and anchorage locations B, respectively; f_c, θ_A , θ_B can be obtained by using the principle of virtual work; , μ_B , μ_C can be calculated through the integration of strain in cross-sections.

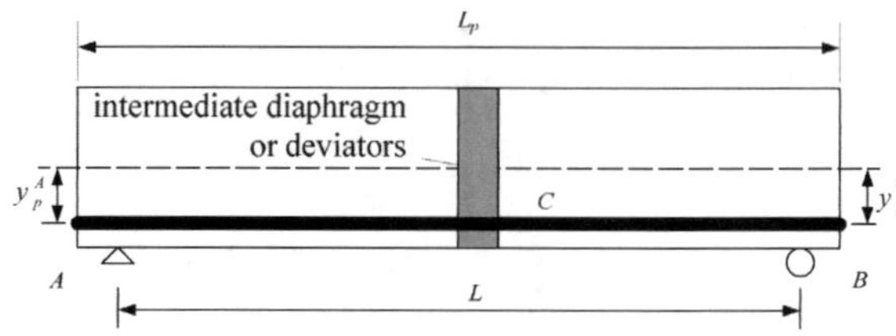

Fig. 2. A beam with straight profiled tendons.

2.5 Calculation method and process

At the beginning of the calculation program, the beam is discretized into n small elements along the axial direction of the beam and assuming that within any element. All of the cross-sections have the same compressive zone height c and curvature φ. Then giving stress increments of externally prestressed tendons $\Delta\sigma_p'$, with (2), (3) and (4), c and φ of any cross-section can be obtained, also the displacements of the locations of the anchorage and deviators, through strain integration and the principle of virtual work. Hereafter, ΔL_p and $\Delta\sigma_p$ can be calculated by using (5) and (6). $\Delta\sigma_p$ and $\Delta\sigma_p'$ will then be compared for iteration until they are close to each other.

Considering material nonlinearity, the all-stage mechanical behavior of simple supported externally PC beam with corrugated steel webs can be calculated by using the nonlinear all-stage-analysis program. Taking account of different loading conditions, external tendons profiles and cross-section shapes, the program can calculate not only the stress and strain in concrete, reinforcement steels and prestressed tendons but also the deflection of the beam for all loading stages from start to failure.

3 Verification of the program

To show the efficiency and the reliability of the program proposed, the program calculated results of one existing experimental beam are compared with its experimental results.

The experimental beam [11], with I-section and single 4 mm thick corrugated steel web, is 400 mm in depth and simply supported over a span of 3000 mm. It is post-tensioned externally by two Φ^s15.24 steel tendons with straight profile and draped at two symmetrically placed deviators at the third-point. The anchorages and two intermediate diaphragm or deviators of the specimen are located 100mm above the soffit. It is subjected to third-point loading. The initial effective prestress of external tendons is 645.9 MPa. Both of the thicknesses of the top concrete flange and bottom concrete flange are 70mm. The material properties are shown in Table 1.

The all-stage curves of load vs. deflection and load vs. external tendon stress are compared with the numerical results in Figure 3 and 4 respectively, showing reasonable agreement. Despite some discrepancy, the calculation results capture the trend of the evolution of external tendon force and mid-span deflection well. The ratio of the program calculated ultimate load with its measured value is 0.92. The ratio of the program calculated ultimate mid-span deflection with its measured value is 0.94. The ratio of the program calculated ultimate external tendon stress increments with its measured value is 0.93. The discrepancy between the program calculated values and the measured is within 10%.

Table 1. Material properties of experimental beam of Harbin Industrial University (unit: MPa).

	Concrete	External tendons	Reinforcement steels	Steel web
Young's modulus(MPa)	31736	196000	210000	205000
Yield strength(MPa)	-	-	330	315
Ultimate strength(MPa)	34.7(cube)	1860	-	-

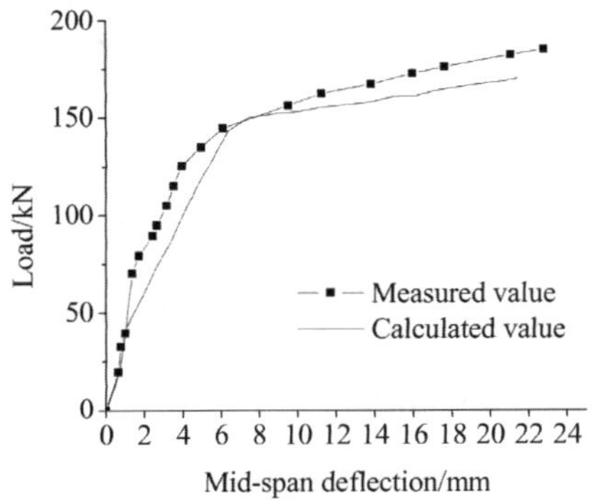

Fig. 3. Measured and calculated mid-span deflection.

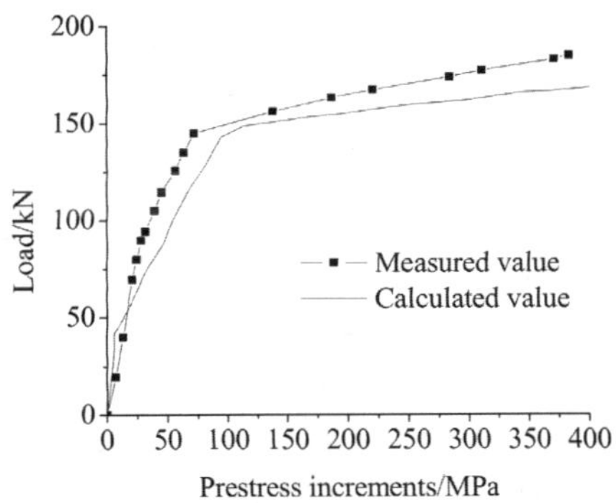

Fig. 4. Measured and calculated stress increments.

4 Parameters analysis

With HIT(Harbin Industrial University) test beam being beam A, varying someone parameter of beam A and keeping other parameters unchanged, beams with different parameters will be got. All of these beams will be calculated to study the influence of important parameters to the bending performance of externally PC beam with corrugated steel webs. The parameters of beam A is listed in Table 2, the value of ultimate mid-span deflection and ultimate load/cracking load are amplified 10 times to their real values.

4.1 Tension steel area

Varying the tension steel area of beam A from 300mm2 to 1500mm2 and keeping other parameters unchanged, four beams named B1, B2, B3, B4 can be got and their calculated results are listed in Table 3. It can be seen from Table 3 that with the increase of tension steel area, the failure of the beam changes from tensile steel pulled off to compressive concrete crushed. The ultimate stress increments of external tendons decrease insignificant at first, then decrease sharply after a value of tension steel area, with which the beam (i.e. beam B3) has the failure of tensile steel pulled off and compressive concrete crushed simultaneously. The ultimate mid-span deflection increases insignificant at first, then reduces sharply after the tension steel area of B3. The ultimate load and the ratio of the ultimate load to the cracking load increase all along.

When tension steel area increases 100%, the ultimate stress increments reduce 7% and 135% before and after the tension steel area value of B3, respectively. The ultimate mid-span deflection increases 2% before the special value and decreases 73% after the special value. Both of the ultimate load and the ratio of the ultimate load to the cracking load increase 21%. The tension steel area has great influence on mechanical properties of externally PC beam with corrugated steel webs.

When tension steel area of beam A changed to the same value with beam B3 and keeping other parameters unchanged, the beam has the failure of tensile steel pulled off and compressive concrete crushed simultaneously. The ultimate mid-span deflection and the ultimate stress increments of externally tendons get to the maximum value. The performance of tensile steel and compressive concrete are both brought into full play and the beam has sufficient ductility with no brittle fracture occurring. The tensile reinforcements can improve the ductility of this beam.

Table 2. Parameters of beam A.

Tension teel area (mm2)	Compression teel area (mm2)	Concrete cube strength (MPa)	Effective prestress (MPa)
566	201	34.7	645.9

Table 3. Calculated values of beams with different tension steel area.

Name of beam	Tension teel area (mm2)	Cracking load (kN)	Ultimate load/ cracking load	Ultimate Load (kN)
B1	300	32.99	4.35	143.39
A	566	32.45	5.32	172.65
B2	700	32.45	5.76	186.76
B3	1270	32.11	7.63	245.10
B4	1500	31.9	8.07	264.82
Name of beam	Ultimate stress increments (MPa)	Ultimate mid-span deflection (mm)	Ultimate strain of tensile reinforcement	Ultimate strain of top edge concrete
B1	447.68	21.68	0.0099	0.0020

A	443.04	22.35	0.0099	0.0022
B2	436.97	22.43	0.0100	0.0023
B3	377.48	22.75	0.0100	0.0030
B4	285.37	18.93	0.0082	0.0030

4.2 Compression steel area

Varying the compression steel area of beam A to 402mm2 and 603mm2 respectively and keeping other parameters unchanged, two beams named C1, C2 can be got and their calculated results are listed in Table 4. It can be seen from Table 4 that with the increase of tension steel area, the ultimate stress increments of external tendons and the ultimate load increase insignificant all along. The ultimate mid-span deflection keeps almost unchanged. The ratio of the ultimate load to the cracking load decreases insignificant all along.

When compression steel area increases 100%, the ultimate stress increments increase 2%. The ultimate load increases 5%. The ratio of the ultimate load to the cracking load decreases 6%. The tension steel area has limited influence on mechanical properties of externally PC beam with corrugated steel webs. The tensile reinforcements can improve the ductility of this beam.

Table 4. Calculated values of beams with different compression steel area.

Name of beam	Compression teel area (mm2)	Cracking load (kN)	Ultimate load/ cracking load	Ultimate load (kN)
A	201	32.45	5.32	172.65
C1	402	41.15	4.61	189.78
C2	603	41.15	4.65	191.44
Name of beam	Ultimate stress increments (MPa)	Ultimate mid-span deflection (mm)	Ultimate strain of tensile reinforcement	Ultimate strain of top edge concrete
A	443.04	22.35	0.0099	0.0022
C1	452.7	22.41	0.0010	0.0021
C2	458.51	22.34	0.0010	0.0016

4.3 Concrete strength

Varying the concrete cube strength of beam A to 60MPa and 80MPa respectively and keeping other parameters unchanged, two beams named D1, D2 can be got and their calculated results are listed in Table 5. It can be seen from Table V that with the increase of concrete cube strength, the ultimate stress increments of external tendons and the ultimate load increase insignificant all along. The ultimate mid-span deflection keeps almost unchanged. The ratio of the ultimate load to the cracking load decreases insignificant all along.

When concrete cube strength increases 100%, the ultimate stress increments increase 19%. The ultimate mid-span deflection increases by 7%. The ultimate load increases 7%. The ratio of the ultimate load to the cracking load decreases 42%. The tension steel area has limited influence on mechanical properties of externally PC beam with corrugated steel webs. The increase of concrete strength will greatly reduce the ductility of the beam.

Table 5. Calculated values of beams with different concrete strength.

Name of beam	Concrete cube Strength (MPa)	Cracking Load (kN)	Ultimate load/ cracking load	Ultimate load (kN)
A	40	32.45	5.32	172.65
D1	60	48.68	3.95	192.16
D2	80	57.80	3.40	196.55
Name of beam	Ultimate stress increments (MPa)	Ultimate mid-spandeflection (mm)	Ultimate strain of tensile reinforcement	Ultimate strain of top edge concrete
A	443.04	22.35	0.0099	0.0022
D1	455.43	22.19	0.0100	0.0019
D2	484.18	22.69	0.0100	0.0017

4.4 Effective prestress

Varying the effective prestress of beam A to 323MPa and 968.9MPa respectively and keeping other parameters unchanged, two beams named E1, E2 can be got and their calculated results are listed in Table 6. It can be seen from Table 6 that with the increase of effective prestress, the ultimate stress increments of external tendons and the ratio of the ultimate load to the cracking load decrease all along. The ultimate mid-span deflection keeps almost unchanged. The ultimate load increases all along.

When the effective prestress increases 100%, the ultimate stress increments decrease 2%. The ultimate mid-span deflection increases by 2%. The ultimate load increases 13%. The ratio of the ultimate load to the cracking load decreases 30%. The effective prestress has some influence on mechanical properties of externally PC beam with corrugated steel webs. The increase of effective prestress will greatly reduce the ductility of the beam.

Table 6. Calculated values of beams with different concrete strength.

Name of beam	Effective prestress (MPa)	Cracking Load (kN)	Ultimate load/ cracking load	Ultimate load (kN)
E1	323.0	20.94	7.88	165.03
A	645.9	32.45	5.32	172.65
E2	968.9	63.00	3.32	209.24
Name of beam	Ultimate stress increments (MPa)	Ultimate mid-span deflection (mm)	Ultimate strain of tensile reinforcement	Ultimate strain of top edge concrete
E1	449.16	22.02	0.0100	0.0020
A	443.04	22.35	0.0099	0.0022
E2	432.36	22.90	0.0100	0.0025

In summary, the effects of these four parameters on the bending performance of externally PC beam with corrugated steel webs in descending order are: tension steel area, concrete strength, effective prestress, compression steel area.

5 Conclusions

(1) Applying the compiled computer program based on moment-curvature to analyze the bending performance of simply supported externally PC beam with corrugated steel webs is feasible.

(2) The tension steel area affected the bending performance of simply supported externally PC beam with corrugated steel webs mostly, followed by the concrete strength and effective prestress. The compression steel area has the minimal impact on the bending performance of this beam.

References

1. M. Cheyrezy and J. Combault, Composite bridges with corrugated steel webs -achievements and prospects,IABSE Symposium on Mixed Structures including New Materials, Brussels, 1990, pp. 479-484.
2. C. Machimdamrong, E. Watanabe and T. Ustunomiya, Analysis of corrugated steel web girders by an efficient beam bending theory, Structural Engineering, 2004, 21(2), pp. 131s-142s.
3. J.F. Bariant, T. Utsunomiya and E. Watanabe, Elasto-plastic analysis of PC girder with corrugated steel web by an efficient beam theory, Structural Engineering, 2006, 23(2), pp. 257s-268s.
4. X.C. Chen, F.T.K. Au, Z.Z. Bai, Z.H. Li and R.J. Jiang, Flexural ductility of reinforced and prestressed concrete sections with corrugated steel webs, Computers and Concrete, 2015, 16(4), pp. 625-642.
5. H.J. LI, S Wan and J.S. Ye, Structure Features of Prestressed Concrete Box girder with Corrugated Steel Webs, Journal of Highway and Transportation Research and Development, 2002, 19(3), pp. 54-57.
6. R Luo and B Edlund, Shear Capacity of Plate Girders with Trapezoidally Corrugated Webs, Thin-walled Structures, 1996, 26(1), pp. 19-44.
7. R. P. Johnson and J. Cafolla, Corrugated webs in plate girders for bridges. Proceedings of the ICE - Structures and Buildings, 1997, 122(2),pp. 157 -164.
8. Y.Q Liu, Steel concre Hybrid Bridge (China Communications Press, Beijing, China, 2005).
9. M. Elgaaly, A. Seshadri and R.W. Hamilton, Bending Strength of Steel Beams with Corrugated Webs, Journal of Structural Engineering, 1997, 123(6), pp. 772-782.
10. J.Y. Song, S.R. ZHANG and Z.L. Wang, Whole Bending Process of Externally Prestressed Composite Beam with Corrugated Steel Webs, Journal of Harbin Institute of Technology University, 2003, 35(5), pp. 590-593.
11. J.Y. Song, Mechanical Properties of Externally Prestressed Composite Beam with Corrugated Steel Webs, Harbin Institute of Technology, Harbin, China, 2003.

Performance of Glazing Materials for Atrium in Hot & Humid Climate

Suhendri[1], *Wahri* Sunanda [2] and *Rachmawan* Budiarto[3]

[1]Institut Teknologi Bandung, Architecture Program, Bandung, Indonesia
[2]Universitas Bangka Belitung, Engineering Physics Department, Pangkalpinang, Indonesia
[3]Universitas Gadjah Mada, Engineering Physics Department, Yogyakarta, Indonesia

Abstract. One comfort issue that considered to be the most important is thermal comfort. Air temperature is the most influential factor, and it is related to solar radiation and building's glazing. Besides thermal comfort, glazing is also needed to provide daylighting. Therefore, due to its relation to thermal comfort and daylighting, the glazing materials selection is crucial. Hence, by focusing on atrium in hot and humid climate, this study is aimed to examine a sufficient material with good thermal and daylighting performance. The results show that argon filled glazing is the most suitable material for atrium glazing in hot and humid climate. Argon filled glass has lowest U-value and total solar transmission than any other glazing materials simulated, but its light transmission is not too low so the daylight delivered still adequate to the standard. Accordingly, U-value, solar heat gain coefficient, and light transmission coefficient are concluded as the determinant factor in selecting glazing material.

1 Introduction

One of comfort issues that considered being the most important is thermal comfort. The annual survey performed in the US, however, shows that 30% of building occupants are dissatisfied with indoor temperature and 80% of complaints about buildings relate to thermal comfort [1]. Even more in hot and humid climate, thermal comfort is a common issue. Air temperature is the most influential factor, and it is related to solar radiation penetrate into a building through building's glazing [2]. Thus, solar radiation, especially in hot and humid climate, has a relation to the thermal comfort inside the building and cooling load of building [3]. Therefore, windows and other openings in building envelope play important role to provide thermal [4].

In addition to thermal comfort issue, openings in building envelope are necessary to provide daylighting. Daylighting is a natural lighting source which provides visual comfort and the delightful of indoor environment, and it has been a primary source of lighting in buildings [5]. Daylighting also has a significant effect on occupant's health as well as occupant's productivity [6]. In brief, light and heat as elements of solar radiation have been important variables that affect the occupant's comfort [7]. Therefore, the design of glazing area of a building and its materials need to reduce solar gains and provide adequate levels of daylight as well [8].

The strategies to deliver daylighting through glazing area in building envelope have been practiced since the beginning of architecture itself. There are two methods to deliver daylight into building, by side-lighting and top-lighting. Top-lighting as one of the strategies did not get attention to be researched as much as side-lighting strategy. Focusing on top-lighting method through atrium in hot and humid climate, this study is aimed to examine a sufficient material with good thermal and daylighting performance. Thus, the study needs to compare different glazing technologies. From the comparison, this study reports the general principles in selecting glazing material for atrium in hot and humid climate.

1.1 Thermal Comfort

Thermal comfort is defined as the condition of mind that expresses satisfaction with the thermal environment [9]. In general, there are two categories of variables that related to thermal comfort, environmental variables, and personal variables. Environmental variable consists of air temperature, mean radiant temperature, air velocity, and relative humidity. The personal variables are metabolic rate due to activity level, clothing insulation [10].

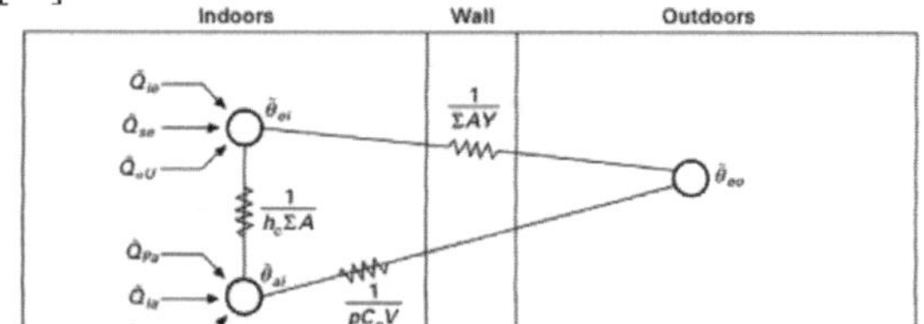

Fig. 1. Nodal network diagram under transient environmental conditions [11].

The relation between indoor air temperature, and thus thermal comfort, to outdoor condition is described adequately using nodal network diagram as shown by Figure 1. The diagram uses three temperature nodes, represent three hypothetical temperatures that interact each other by transferring heat.

- Outdoors condition is represented by sol-air temperature (θ_{eo}). It is affected by shortwave radiation, longwave radiation, and convective heat exchanges.
- Indoor condition is represented by two nodes:
  - Environmental temperature, affected by conductive gains through the fabric, internal gains, and glazing solar gains.
  - Mean air temperature, affected by plant load (mechanical heating/cooling), convective internal gains, and glazing solar gains.

Nodal diagram network gives better understanding on the role of solar radiation in determining indoor conditions. Yet, solar radiation is an important feature in lighting design as explained in the Daylighting Section below. In this manner solar radiation is recognised to be a parameter that is needed and avoided at the same time.

1.2 Daylighting

Visual comfort is a multidimensional subject, and thus not sufficient to use only illumination levels for its evaluation. British Standard has established daylight factor and average daylight factor as a measurement in daylighting. Daylight factor, as mentioned in BS 6100-7:2008, is ratio of illuminance at a point on a given plane due to light received from a sky of known or assumed luminance distribution, to illuminance on a horizontal plane due to an unobstructed hemisphere of this sky. While average daylight factor is defined as ratio of total daylight flux incident on a reference area to total area of reference area, expressed as a percentage of outdoor illuminance on a horizontal plane due to an unobstructed hemisphere of sky of assumed or known luminance distribution [12].

Illuminance and daylight factor is controlled by glazing design. The chosen material for glazing plays an important role in delivering good daylighting. With two parameters, daylight factor and illuminance, performance of the glazing materials are compared. British Standard gives a minimum daylight factor requirement for general building at least 2% [12]. Meanwhile, the recommended illuminance levels needed for atrium is minimum 50-200 [13].

2 Methodology

The case study building is Marina Mandarin Hotel in Singapore. Its location is on the north side of Marina Bay. The architect of the building is John Portman & Associates. Marina Mandarin is a 5-star hotel located in Marina Bay, and is part of the Marina Square mixed-use complex. Total area of the Marina Square mixed-use complex is 4 million square foot, including three convention hotels, a retail mall, and an office building surrounded by plazas and gardens [14].

One of the main features of the hotel structure is its magnificent atrium. Glazing material of the atrium is clear glass. However, due to Singapore's hot climate, solar heat still penetrates through the atrium and the clear glass cannot perform well in such condition. It makes intense heat gain to the interior [15]. The suitable material for the building is examined by comparing the performance of original glazing material with other materials.

Singapore itself is located in tropical zone, with warm humid climate classification. Average maximum temperature in Singapore is 32.18°C and its average minimum temperature is 23.27°C. Singapore has average maximum relative humidity of 80.67 % and average minimum relative humidity of 72.92%. The main problem in this type of climatic zone is overheating. The overheating is provoked by high humidity, restricting the evaporation potential. The diurnal temperature variation is small. The primary concern in Singapore's climate is high solar altitude and radiation. Therefore, shading devise becomes a meaningful part for the building.

Fig. 2. Case building view [14].

Table 1. Properties of Glazing Technologies Used in the Simulation.(Source: [4])

	Solar transmission	**Direct solar transmission**	**Light transmission**	**U-value**
1. Multilayer glazing				
Single Clear Glazing	0.810	0.775	0.881	6.121
Double Clear Glazing	0.693	0.604	0.781	3.157
Triple Clear Glazing	0.675	0.595	0.738	2.166
2. Low-emittance glass				
Single Low-e Glass	0.710	0.680	0.811	4.233
Double Low-e Glass	0.595	0.538	0.769	2.440
3. Electrochromic glass				
Double Electrochromic Glass	0.638	0.545	0.727	1.772
4. Gas-filled glass				
Double Argon-Filled Glass	0.564	0.745	0.745	1.499

In order to achieve the aim and objectives, the study employs two phases or steps. First of all, some glazing technologies that are usually used for glazing will be simulated using DesignBuilder software. Then, comparison is done to the simulation result. The technologies are, multilayer glazing, low-e glass, glass filled with absorbing gases, and electrochromic glass. Secondly, the parameters in selecting the glazing technologies for atrium in building method are derived from the simulation results (Table 1).

3 Results & Discussion

3.1 Multilayer Glazing

3.1.1 Analysis of Thermal Environment

Performance of the multilayer glazing is observed by using temperature in the atrium hall below the glazing. There are three temperatures to be compared, air, radiant, and operative temperature. The comparison can be seen in Figure 1-3. From the graph, it can be seen that all of the glazing still can reduce the outside temperature. However, it also shows that the more glazing layer causes warmer temperature. It indicates that adding more glazing layer perhaps not an appropriate strategy for atrium in hot and humid climate.

Supporting evidence for the findings in Figure 4 can be explained by using the comparison of glazing heat losses data. From the data shown in Figure 4, heat losses from the atrium glazing are getting lower, reciprocal to the increasing number of glazing layer. It is because every glazing layer added to the glazing, reduces the U-value of the glass. Thus, heat transfer occurs in a lower rate. Hence, there is more heat trapped and it increases the temperature.

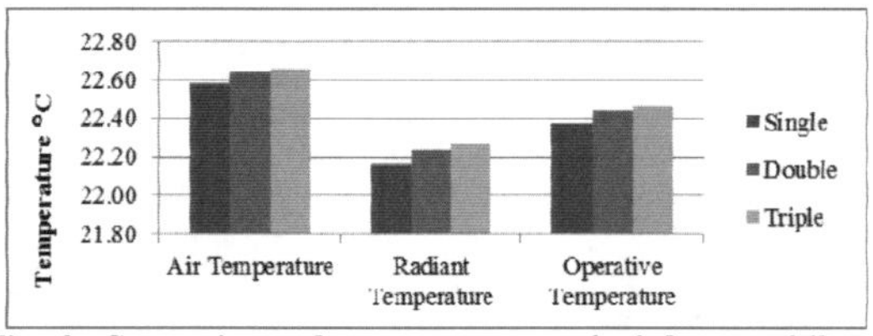

Fig. 3. Comparison of temperatures resulted from multilayer clear glass.

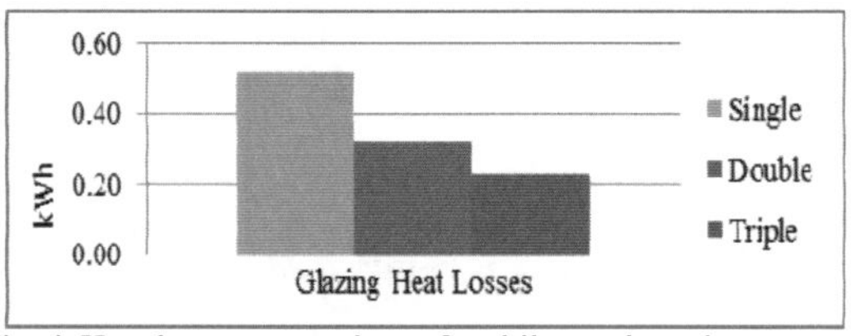

Fig. 4. Heat losses comparison of multilayer clear glass.

3.1.2 Analysis of Daylighting

By using the contour images from DesignBuilder, daylighting in the atrium below the glazing. Figure 5 shows the results comparison. According to illuminance minimum requirement for atrium building, the light should be between 50-200 lux. From the figure it can be seen that all of the three glazing materials deliver appropriate light. Generally, single clear glazing gives brighter daylight, although it is not the most uniform daylight distribution. It is due to light transmission of each glazing materials. More glazing layer means less light transmitted. Hence, single glazing passes more light.

Overall, triple glazing has the lowest average daylight factor. However the values are quite similar, so the distribution of the light is the key to determine the preferred glazing. From Figure 5, it can be noticed that double glazing performs a better light distribution. There is also no significant glare for double glazing material.

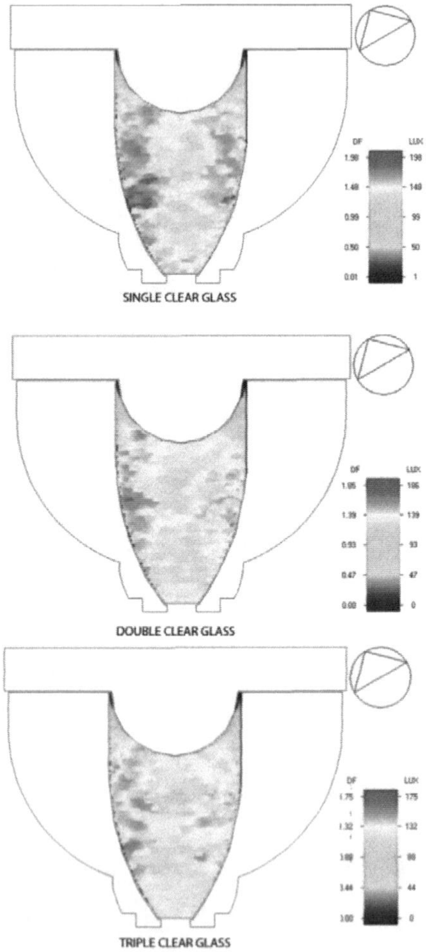

Fig. 5. Contour image of daylight delivered by multilayer clear glass.

3.2 Clear Glass Compare to Other Glazing Technologies

3.2.1 Analysis of Thermal Environment

General comparison of three coating technologies with the clear glass is shown by Figure 6. Indoor temperatures seem to be similar for all of the glazing technologies.

 https://doi.org/10.1051/matecconf/201818602007

Figure 6 shows that Low-e glazing technology has no impact to the indoor temperature of the atrium. More, the heat gains/losses of Low-e glazing also has no difference with the clear glass as shown by Figure 7. This result is not as expected, since the U-value of these two materials is different. Low-e glass is expected to reduce the indoor temperature, but the prediction does not give the expected result.

The similar interpretation with the multilayer glazing can be used to explain the phenomena. The heat is trapped inside the atrium due to low U-value. However, since the direct total solar transmission through low-e glazing is much lower than clear glass, the effect of the heat trapped is abolished by small amount of heat from solar radiation.

Furthermore, low-e argon filled glazing and electrochromic glazing reduce the temperature in the atrium. Figure 6 shows the temperature reduction occurs at the same value. Glazing losses data supports the temperature data. Low-e argon filled glass and electrochromic glass gain less heat than clear glass and low-e glass. However, the temperature data shows the same reduction value although the glazing losses data shows low-e argon filled glass performs better.

The analysis can be aided by looking at properties of the two materials. The two materials, especially argon filled glass, have the lowest U-value. They also have low SHGC value. This means both materials reduce the temperature by keeping heat from outside to not entering the building instead of losing the heat from inside the building. Moreover, argon filled glass –which has lower U-value and total solar transmission than electrochromic glass– performs best for providing good thermal environment. However, this statement should be confirmed by examining the daylight performance of each glazing technology in the following discussion.

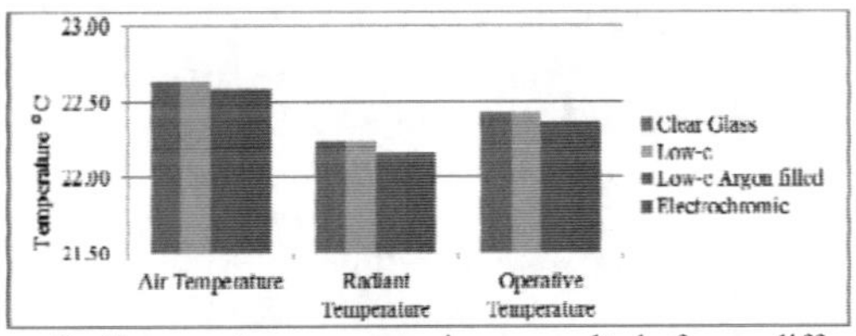

Fig. 6. Temperatures comparison resulted from different glazing technologies.

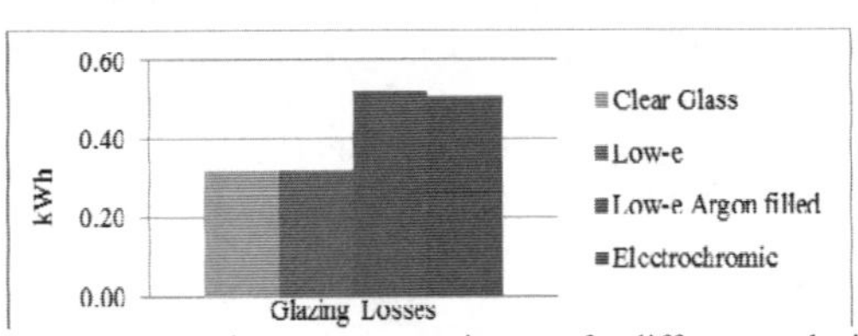

Fig. 7. Heat losses comparison of different glazing technologies.

3.2.2 Analysis of Daylighting

Similar with multilayer glazing, overall daylight performance meets the requirement of daylight levels from CHANNEL (2011). The average illuminances for the three materials range between 50-200 lux. Figure 8 compares daylight performance of clear glass and low-e glass. It shows that low-e glass delivers dimmer daylight. Besides, the distribution of the light is not as uniform as clear glass. The sunlight seems to be scattered by the low-e coating installed to the glazing and, it makes the light distributed inconsistently. However, its goal to reduce the incoming solar radiation –and thus reduce the temperature– is still being achieved.

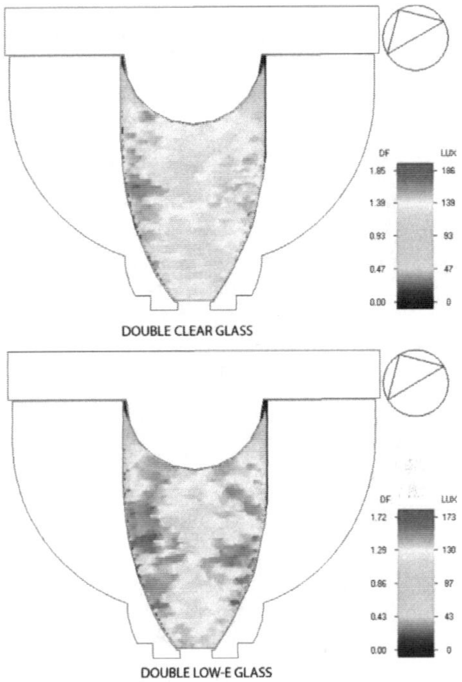

Fig. 8. Clear glass and low-e glass daylight comparison.

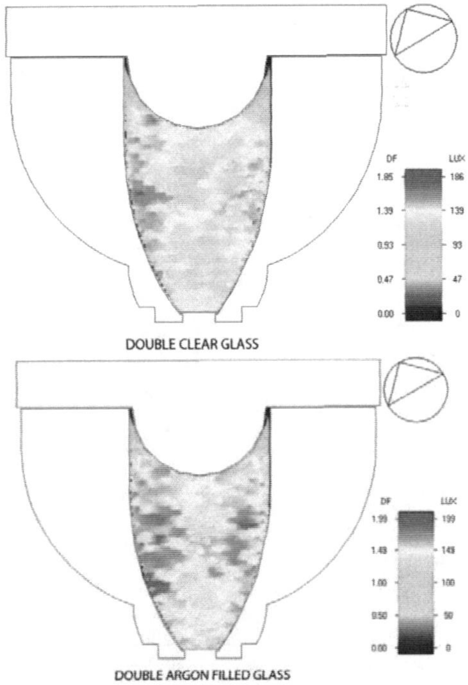

Fig. 9. Clear glass and argon-filled glass daylight comparison.

 https://doi.org/10.1051/matecconf/201818602007

DesignBuilder simulation for argon filled glass gives different result from low-e glazing. As shown by Figure 9, argon filled glass brings more light than clear glass and low-e glass. But, the distribution of the daylight is not as homogeneous as clear glass. However, the light delivered is still in the standard range, without any significant glare. Compare to low-e glass, daylight from argon filled glass is more preferred because it is brighter and more uniform.

Comparing electrochromic glass with the clear glass, it shows that electrochromic glass delivers brighter light. At few locations in the atrium, the daylight illuminance is even more than the range of preferred illuminance for atrium, as seen in Figure 10. Among the three glazing materials, electrochromic gives the most uniform daylight distribution. But, its illuminance is a disadvantage for electrochromic to be considered as the most appropriate material. At last, argon filled glazing performs better than the other glazing technologies. It delivers sufficient daylighting and can reduce the temperature. Furthermore, analysis of its properties supports the fact that argon filled glass is the most appropriate glazing technology.

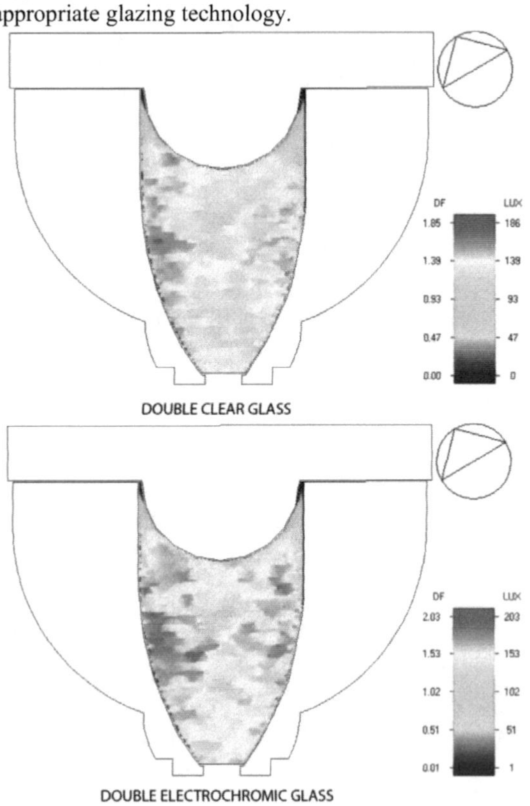

Fig. 10. Clear glass and electrochromic glass daylight comparison.

4 Conclusion

In terms of thermal performance, single glazing is in the first-rate. On the other hand, double clear glazing technology performs better than single and triple glazing in terms of daylighting performance. More glazing makes the temperature warmer because it traps the heat from inside the building. Though, this effect of increasing indoor temperature is negligible since it happens in small differences. Further, more glazing is needed to reduce light transmitted through it. Hence, triple glazing gives dimmer light than the others, although double glazing has the most uniform daylight distribution.

In addition, the analysis of performance of the other glazing technologies–low-e glass; argon filled glass; and electrochromic glass–brings out a fact that argon filled glazing technologies gain less heat than clear glass and low-e glass. argon filled glass performs better for providing good thermal environment. Argon filled glazing technology is the most appropriate for both thermal comfort and daylighting. Hence, argon filled glazing is concluded to be the most suitable material for atrium glazing in hot and humid climate.

Acknowledgment

We gratefully acknowledge the funding from USAID through the SHERA program - Centre for Development of Sustainable Region (CDSR).

References

1. P. Smealle, *Building Occupant and Customer Satisfaction Survey Results for 2001*, (2003)
2. M. R. Hall, *Materials for Energy Efficiency and Thermal Comfort in Buildings*, (2010)
3. S. Li, K. Zhong, Y. Zhou, and X. Zhang, *Comparative Study on The Dynamic Heat Transfer Characteristics of PCM-filled Glass Window and Hollow Glass Window*, J. Energy Build., **85**, 483–492, (2014)
4. E. Cuce and S. B. Riffat, *A state-of-the-art review on innovative glazing technologies*, Renew. Sustain. Energy Rev., **41**, 695–714, (2015)
5. C.-S. Kim and S.-J. Chung, *Daylighting simulation as an architectural design process in museums installed with toplights*, Build. Environ., **46**, 210–222, (2011)
6. C. Meek and K. Van Den Wymelenberg, *Daylighting and Integrated Lighting Design*, (2014)
7. K. M. Al-Obaidi, M. Ismail, and A. M. A. Rahman, *A study of the impact of environmental loads that penetrate a passive skylight roofing system in Malaysian buildings*, Front. Archit. Res., **3**, 178–191, (2014)
8. B. Ercan and S. T. Elias-Ozkan, *Performance-based parametric design explorations: A method for generating appropriate building components, Des. Stud.*, **38**, 33–53, (2015)
9. ASHRAE, ASHARE *Standard 55. Thermal Environmental Conditions for Human Occupancy*, 1–52, (2003)

10. D. Kalz and J. *Pfafferott, Thermal comfort and energy-efficient cooling of nonresidential buildings*, (2014)
11. M. R. Hall and D. Allinson, *Heat and Mass Transport Processes in Building Materials, in Materials for Energy Efficiency and Thermal Comfort in Buildings*, (2010)
12. BRE, British Standard, (2008).
13. CHANNEL, *Lighting & Safety Systems*, (2011)
14. ARCHITECTURE SG., *John Portman's Singapore*, (2014), https://architecturesg.wordpress.com/2014/12/08/john-portmans-singapore/
15. Eastman Chemical Company, *Architecture Window Films*, (2013), www.iquefilms.com

https://doi.org/10.1051/matecconf/201818602008

Wind Fragility for Sign Structure in Korea with Chemical Anchor Connection

Viriyavudh Sim[1], *SeongDo* Kim[1], *WooYoung* Jung[1]

[1]Gangneung-Wonju National University, Department of Civil Engineering, 25457 Gangneung, Republic of Korea

Abstract. Damage to domestic facilities by strong winds and typhoon is on the rise in recent year. Typical sign structure used in Korea is found on the edge of concrete building, which could be easily affected by strong winds. Therefore, this study focused on sign structure among various vulnerable facilities. The evaluation of wind fragility for sign structure was carried out considering the failure of anchor at their connection to the concrete building. Moreover, pull-out and shear tests were performed to determine the resistance capacity of anchor used in this study. Monte Carlo Simulation method was used to generate random wind loads on sign structure. Additionally, fragility parameters were determined based on the failure criteria of connection anchor. Results show that diameter of the installed anchor was the dominant factor affecting the performance of sign structure subjected to strong wind.

1 Introduction

Per Korea Typhoon White Book from Korea National Typhoon Center 2011 [1], typhoon disaster in Korea in recent year has been on the rise. The risk due to typhoon causes devastating loss on economy and human life. Therefore, the risk assessment for structure subject to strong wind loads is required. In fact, probabilistic risk assessment (PRA) has emerged as an increasingly popular analysis tool, especially in the last decade. For instance, Lee and Rosowsky [2] and Ellingwood, Rosowsky, Li and Kim [3] analyse fragility curve for wood frame structure subjected to lateral wind loads by means of Monte Carlo Simulation (MCS) method, which is a part of analytical fragility development. The method was based on statistical wind load parameters determined by Ellingwood and Tekie [4] by means of Delphi questionnaire with expert of this field. Moreover, by combining the statistical parameters with nominal wind load parameters in standard design code from American Society of Civil Engineering (ASCE 7-10) [5], it is applicable to variety of structure types.

PRA is a systematic and comprehensive methodology to evaluate risks associated with every life-cycle aspect of structural and non-structural component. Recent disasters around the world have highlighted the necessity of risk assessment for all types of structure, especially those located in high density region or metropolitan area, which prone to cause higher economic losses and social disruption. Advertisement sign structures in South Korea are those mostly found in crowded region and mostly constructed in the form of attachment to the side or on the roof of concrete structure. Most published studies use regression technique with post-disaster investigations or historical recorded data to develop fragility curves [6]. However, this approach cannot be used to develop fragility curves for structure in South Korea because of the limitation of post-disaster and historical data [7].

Fig. 1. Failure of sign structure.

In this paper, a probabilistic approach was presented to predict the wind-induced damage of sign structure. This approach used Monte Carlo Simulation method to generate damage information for structural component. Additionally, considering the experts' opinions of related companies, the damage of sign structure frequently occurs at the attachment surface, which can be seen in Figure 1; thus, the connection anchors were selected as vulnerable member in this study. Subsequently, the simulation compared probabilistic wind loads and resistance strength of structural component based on experimental test data of chemical anchor frequently used in South Korea. Finally, the wind fragility of the chemical anchor was evaluated. Multiple types of chemical anchor were considered based on their length and diameter.

2 Performance evaluation of chemical anchor

Pull-out and shear test was performed to determine resistance capacity of chemical anchor. The experiment was performed with four types of anchor based on their dimension. The dimensions of the anchors were selected from the most commonly used in the market with lengths of *50 mm* and *100 mm* and diameters of *10 mm* and *12 mm*. Chemical anchor was shown in Figure 2. The experiment was carried out according to: penetration depth and anchor diameter.

Fig. 2. Chemical anchor.

The performance of the chemical anchor was measured three times per experimental variable and the mean value was obtained for each experimental parameter. The preparation of concrete block and experiment procedure was shown in Figure 3. In Figure 3(a)-(c), the concrete bock was constructed and anchor hole was drilled and clean up for installation of anchor. Then, the chemical anchor in Figure 2 was installed. Finally, the pull-out and shear tests were performed as shown in Figure 3(d).

(a). Construction of concrete specimen

(b). Hole drilling for anchor installation

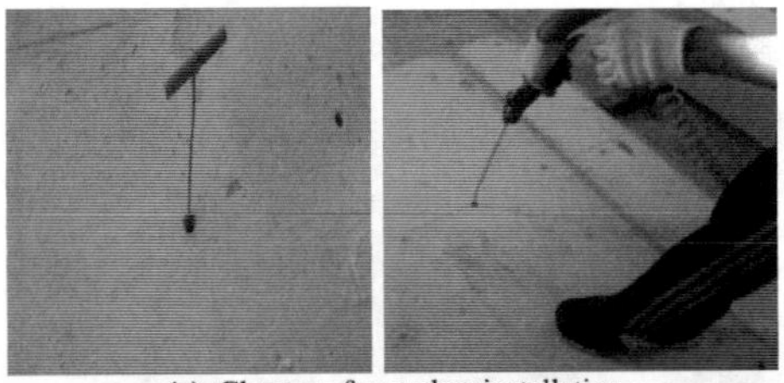

(c). Clean up for anchor installation

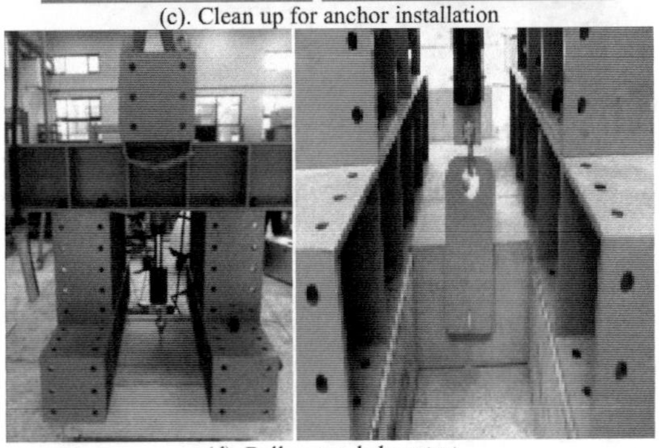

(d). Pull-out and shear test

Fig. 3. Experimental procedure.

Results of the pull-out and shear tests were shown in Figure 4 and Figure 5, respectively. In Figure 4, (a) the failure due to anchor pull-out and (b) the force-displacement curves for three tests were shown. Likewise, in Figure 5, the failure form and force-displacement curves were shown for shear test. Additionally, Table 1 and Table 2 show the results of displacements and forces for all four types of anchor selected for this study.

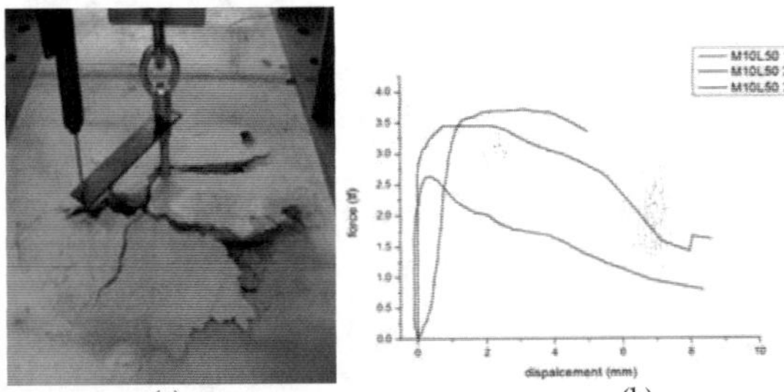

(a) (b)

Fig. 4. Pull-out test results.

Table 1. Chemical anchor pull-out test results.

No.	Anchor	Displacement (mm)	P (ton)
1	M10L50	3.852	3.367
2	M10L100	3.473	3.907
3	M12L50	4.514	4.623
4	M12L100	4.627	9.247

 https://doi.org/10.1051/matecconf/201818602008

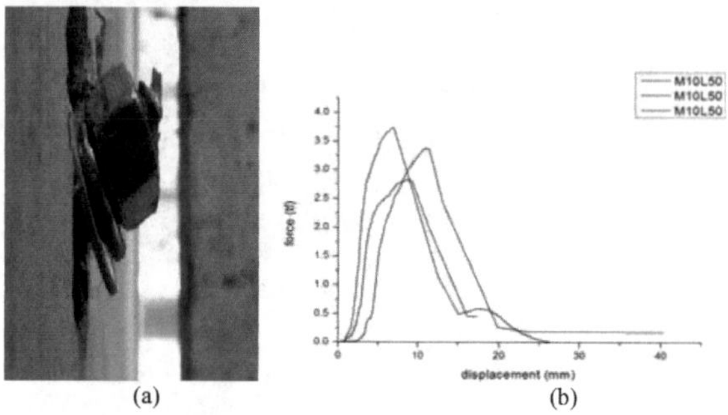

(a) (b)

Fig. 5. Shear test results.

Table 2. Chemical anchor shear test results.

No.	Anchor	Displacement (mm)	P (ton)
1	M10L50	10.995	3.317
2	M10L100	16.528	3.657
3	M12L50	8.631	5.421
4	M12L100	11.105	5.977

In both experiments, the strength of anchor bolts was measured using a hydraulic double acting cylinder (TDC-3030) and a load-controllable server, and the displacement values were also measured by linear variable differential transformer (LVDT). The resistance capacity was estimated based on the experimental results. Finally, we derived and evaluated the wind fragility based on the performance of these anchors.

3 Structure model setting

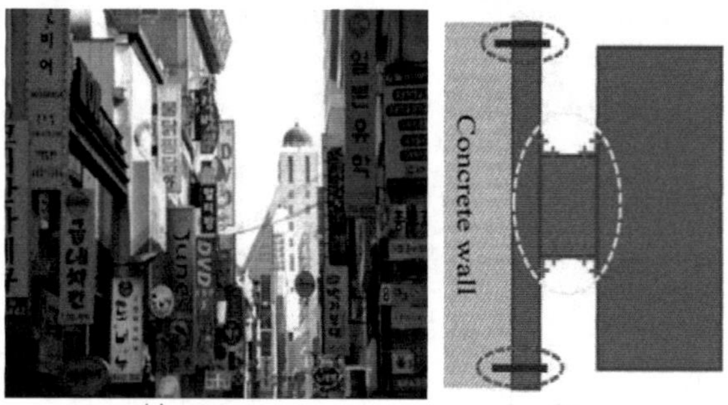

(a) (b)

Fig. 6. Typical sign structure in Korea.

In South Korea, advertisement is commonly placed on the sign structure connected to the side of concrete building in the crowded area as can be seen in Figure 6(a). More specifically, the sign structure model in this study was set up as shown in Figure 6(b) for wind fragility evaluation. The advertisement panel's dimension was $0.8\ m \times 3.2\ m$. In this diagram, the panel was connected to the steel plate through the rigid connection box with 8 screws and the steel plate connected to concrete wall with 4 anchors. Failure modes of the sign structure due to strong wind include connection angle dropout and connection anchor failure; however, in this study, the connection box was assumed to be perfectly rigid, only the anchor failure was considered.

4 Evaluation of sign structure wind fragility

4.1. Wind loads

ASCE 7 (2010) defines two types of structural elements subjected to wind load: (1) main wind-force resisting systems (MWFRS), and (2) components and cladding (C&C). The wind load acting on solid freestanding sign structures shall be determined with provision for MWFR in ASCE 7 (2010) with the following:

$$F = q_h G C_f A_s \qquad (1)$$

where q_h = velocity pressure evaluated at height h (top edge of sign panel), G = gust-effect factor, C_f = net force coefficient, and A_s = gross area of the solid freestanding sign in (m^2). The velocity pressure evaluated at height z in ASCE 7 (2010) is given by:

$$q_z = 0.613 K_z K_{zt} K_d V^2 \qquad (2)$$

where q_h is equivalent to q_z at height h, K_z = velocity pressure exposure factor, K_{zt} = topographic factor, K_d = wind directionality factor, and V = basic wind speed in (m/s) (3-second gust wind speed at $10\ m$ and in open terrain).

Summary of statistical wind load parameters used in this study was shown in Table 3. Normal distribution of parameters K_z, K_d, and G were obtained from previous Delphi study of wind parameters by [2] and [4].

Table 3. Summary of statistical wind load parameters.

Parameter	Category	Mean	SD	CDF
K_z	Exposure B	0.89	0.17	Normal
	Exposure C	1.08	0.15	Normal
	Exposure D	1.28	0.18	Normal
K_d	MWFRS	0.89	0.14	Normal
G	Exposure B	0.77	0.09	Normal
	Exposure C	0.83	0.10	Normal
	Exposure D	0.83	0.07	Normal

C_f	Deterministic (1.93)
K_{zt}	Deterministic (1.00)

4.2 Probability of failure of anchor bolt

Failure of sign structure due to wind loads occurs when force generated by wind exceeds the resistance capacity of connection anchor. Thus, the failure criteria for anchor bolts can be written in term of the basic (random) variables as:

$$g(x) = R - F \tag{3}$$

where R = resistance capacity of the anchor bolt, F = wind loads acting on sign panel that apply tension and shear loading to anchor bolts. Failure of anchor bolts can be defined as a condition where $g(x) < 0$. The wind loads, and hence the probability, is a function of the basic wind speed (V) squared, see (1) and (2).

Monte Carlo Simulation method was used to generate random wind loads (F) based on their statistical distribution in Table 3. For each wind speed, MCS generated random velocity exposure factor, wind directionality factor, and force coefficients by sampling from their normal distribution function parameters. Subsequently, by comparing wind loads demand with anchor resistance capacity, the failure of anchor could be determined. These comparisons were repeated *10,000 times* to determine the probability of failure at each wind speed. Furthermore, by repeating this procedure for the next increment of wind speed, the probability of failure for anchor could be determined for all wind speeds.

4.3 Wind fragility modelling of anchor bolt

The fragility parameters of a structural system commonly modelled using a lognormal cumulative distribution function as the following [8]:

$$Fr(V) = \Phi\left[\frac{\ln(x) - \mu_R}{\sigma_R}\right] \tag{4}$$

in which $\Phi(\cdot)$ = standard normal cumulative distribution function, μ_R = logarithmic median of capacity R (in units that are dimensionally consistent with demand), and σ_R = logarithmic standard deviation of capacity R.

5 Results and discussions

Wind fragility parameters μ and σ were determined by maximum likelihood estimation (MLE) method [9]. Figure 7-9 show the wind fragility of chemical anchor installed in sign structure in region with wind exposure category B. Anchor No. 4 (M12L100) showed the highest resistance to wind loads, where their median wind speed had the highest value, i.e. *103 m/s*. Then, it was followed by anchor No. 3 (M12L50), No. 2 (M10L100) and No. 1 (M10L50) with the median wind speed *93 m/s*, *78 m/s* and *74 m/s*, respectively. Accordingly, there was a minor difference between median wind speed of anchor with the same diameter, i.e. *10 m/s* for *12 mm* anchor and *4 m/s* for *10 mm* anchor. However, for anchor with the same length, a more significant difference between their median wind speed was shown. For anchors with *50 mm* length, the difference between their median wind speed were *19 m/s*; anchor with *100 mm* length shown a difference of *25 m/s*.

This trend was also shown in the other two wind exposure categories. Therefore, it could be concluded that the difference of probability of failure for anchor with the same diameter (*10 mm* and *12 mm*) was less than the difference between anchor with the same length (*50 mm* and *100 mm*). Thus, the diameter of chemical anchor installed for the connection of sign structure and concrete wall can significantly affect the performance of sign structure system.

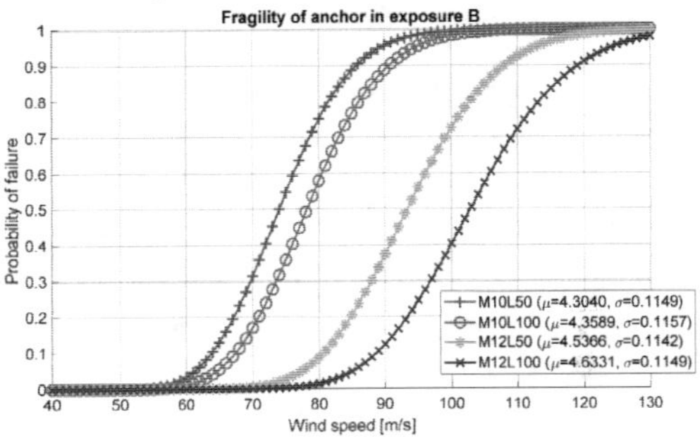

Fig. 7. Wind fragility of chemical anchor in sign structure in wind exposure B.

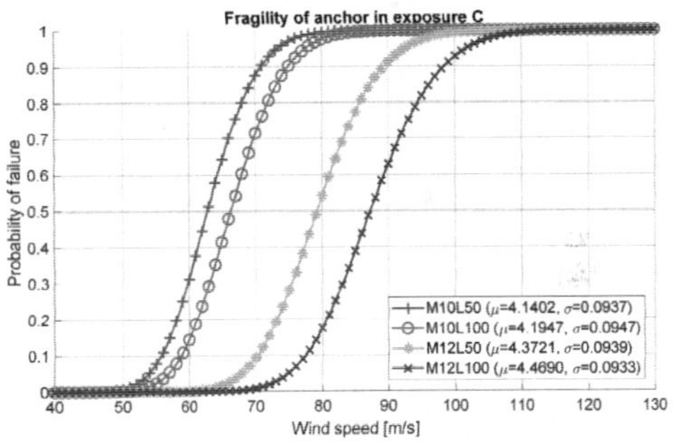

Fig. 8. Wind fragility of chemical anchor in sign structure in wind exposure C.

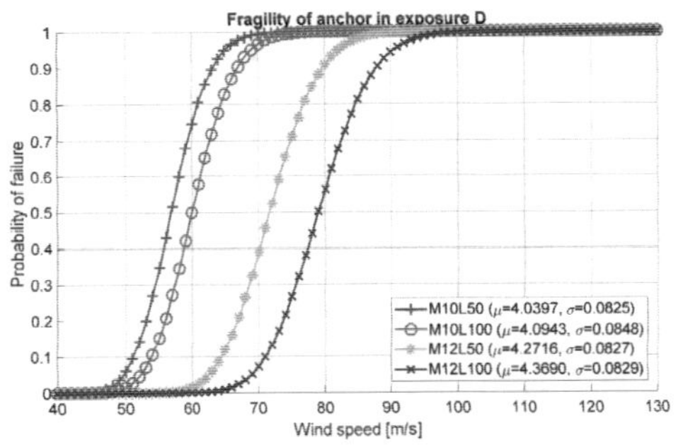

Fig. 9. Wind fragility of chemical anchor in sign structure in wind exposure D.

6 Summary and conclusions

Performance of advertisement sign structure was evaluated based on the performance of their connection anchor to the concrete wall. Probability of failure for sign structure were determined analytically with Monte Carlo Simulation method. Moreover, fragility parameters were fitted to lognormal cumulative distribution function by using the maximum likelihood estimation method. Results were shown for anchor installed in sign structure in region with wind exposure B, C and D. Additionally, the difference between median wind speed of each anchor had also been discussed.

The results illustrated that anchor No. 4 (M12L100) shown the highest resistance; while, anchor No. 1 (M10L50) had the lowest resistance capacity. The differences between median wind speed of each anchor shown the significant effect of anchor diameter on the performance of sign structure exposed to strong wind. It could be concluded that the diameter of chemical anchor was more significant than their length in application where the chemical anchor was used in connection between sign structure and concrete wall.

In conclusion, usage of this methodology could lead to a more predictable structure performance and could facilitate the introduction of performance-base design guidelines for sign structure in high wind regions. Fragilities such as those presented here also can convolve with wind hazard curves to develop a risk assessment tool, which can evaluate the potential impact of a natural hazard in public planning and mitigate the consequent economic losses and social disruption.

Acknowledgment

This research was supported by a grant [MOIS-DP-2015-05] through the Disaster and Safety Management Institute funded by Ministry of the Interior and Safety of Korean government.

References

1. National Typhoon Center, *Typhoon White Book 2011*, Publication Registration Number: 11-1360016-000001-01, (2011)
2. K. H. Lee, D. V. Rosowsky, *Fragility curves for woodframe structures subjected to lateral wind loads*, Wind Struct., **9**, 217-230 (2006)
3. B. R. Ellingwood, D. V. Rosowsky, Y. Li, J. H. Kim, *Fragility assessment of light-frame wood construction subjected to wind and earthquake hazards*, J. Struct. E, **130**, 1921-1930 (2004)
4. B. R. Ellingwood, and P. B. Tekie, *Wind load statistics for probability-based structural design*, J. Struct. E, **125**, 453-463 (1999)
5. American Society of Civil Engineers, *Minimum design loads for buildings and other structures*, ASCE, **7** (2010)
6. A. D. Cope, *Predicting the vulnerability of typical residential buildings to hurricane damage*, Doctoral dis., University of Florida (2004)
7. H. Ham, S. Lee, H. Kim, *Development of typhoon fragility for industrial buildings*, Proc. 7th Asia-Pacific Conf. Wind E, Taipei (2009)
8. K. Porter, *Beginner's guide to fragility, vulnerability, and risk*, Ency. Eq. E, 235–260 (2015)
9. J. W. Baker, *Efficient analytical fragility function fitting using dynamic structural analysis*, Eq. Spect., 31, 579–599 (2015)

Effect of Ti Addition to Age Hardening Response of Al-10Zn-6Mg-xTi Alloy Produced by Squeeze Casting

Dwi Ayu Nurcahyaningsih[1], *Risly* Wijanarko[1], *Irene* Angela[1], and *Bondan* Tiara Sofyan[1]

[1]Department of Metallurgy and Materials Engineering, Faculty of Engineering, Universitas of Indonesia, Depok 16424, West Java, Indonesia

Abstract. This research focused on investigating the effects of Ti addition on the age hardening response of Al 7xxx alloy for Organic Rankine Cycle (ORC) turbine impeller application in power plant generators. Al-10Zn-6Mg wt. % alloys were produced by squeeze casting with 0.02, 0.05, and 0.25 wt. % Ti addition. As-cast samples were homogenized at 400 °C for 4 h. Solution treatment was conducted at 440 °C for 1 h, followed by quenching and ageing at 130 °C for 200 h. Age hardening result was observed using Rockwell B hardness measurement. Other characterizations included impact testing, STA, optical microscopy, and SEM-EDS. Results showed that the addition **of** Ti in all content variations increased the as-cast hardness due to the diminution of secondary dendrite arm spacing (SDAS) values of the alloy. Ageing at 130 °C strengthened the alloys, however the addition of Ti was not found to affect neither peak hardness nor impact values of the alloy. Identities of second phases formed during solidification were found to be T ($Mg_{32}(Al,Zn)_{49}$), β (Al_8Mg_5), and $TiAl_3$, while precipitates produced during ageing were GP Zone, η', and η ($MgZn_2$). **Keywords** – ageing, Al-Zn-Mg, grain refining, $MgZn_2$, Ti addition.

1 Introduction

Al 7xxx alloys with its Zn and Mg alloying content are commonly used for high temperature service. The alloy is currently developed for turbine impeller application in novel Organic Rankine Cycle/ORC power generating system considering its lightweight, strong, and durable properties [1] that would contribute positively to the overall system efficiency.

Grain refining agents – e.g. Ti, Cr, Zr, or other rare earth elements are used in order to further strengthen the alloy. Kurniawan [2] founded that 0.067, 0.081, and 0.115 wt. % Ti content in AC4B alloy all resulted in increased strength and UTS value of the alloy. Addition of 0.050 and 0.072 wt. % Ti was also evident to be able to optimize the hardness of AC4B alloy by 2.80 % and 7.17 %, along with optimizing its tensile strength by 4.40 % and 5.10 % compared to those without Ti addition. The strengthening mechanism was a result of primary α-Al grain refining phenomenon followed by a decrease in dendrite arm spacing/DAS value [3]. Heterogeneous nucleation and grain boundary strengthening effect were also found after 0.058, 0.073, and 0.104 wt. % Ti addition to both squeeze-cast and aged Al-9Zn-6Mg-2Si wt. % alloy with ZrO_2 reinforcement [4].

Sigworth and Kuhn [5] showed that Ti contributed to the grain refining mechanism by rising Al solidification temperature from 660.1 to 665 °C, at which $TiAl_3$ particles formed and initiated the cooling of Al melt to primary α-Al phase. $TiAl_3$ particles induced a heterogeneous nucleation of α-Al, hence a greater number of present $TiAl_3$ would refine the overall grain size. A research by Gao, et. al. [6] displayed an agreement to the statement, in which the addition of 0.2 wt. % Ti to Al-6Zn-2Mg-2Cu-0.2Zr wt. % alloy resulted in smaller α-Al grains in comparison to alloy samples without Ti addition. This is, however, limited to a certain time range during the cooling process before a *fading* event – submerging $TiAl_3$ intermetallic particles to the bottom of crucibles due to a prolonged holding time – occurs. On the occasion where intermetallic nuclei were unable to form, Ti would serve as a growth restricting agent [7]. The grain refining effect, by all mechanisms, will result in higher strength of the Al-alloy in accordance to Hall-Petch equation.

Further strengthening of Al 7xxx alloy is generally done through age hardening/T6 heat treatment method which consists of solution treatment, quenching, and ageing steps, producing precipitates of new phases that render dislocation movement. Possible identities of formed phases include GP Zone [8], η ($MgZn_2$), η' [9], β", $MgSi_2$ [10], cubic T' ($Mg_{32}(Al,Zn)_{49}$), and equilibrium T [11]. Reportedly, the addition of Ti was able to accelerate the precipitation forming in experimental ageing of Al-6Zn-2Mg-2.5Cu-0.2Zr-0.2Ti wt. % at 120 °C [6] and prevent grain boundary segregation from occurring during air cooling [9].

The above literature survey shows contributions of Ti alloying element to both as-cast and after age hardening properties of Al 7xxx alloy. Thus, the present work attempts to analyze the effects of Ti addition at varying contents to the microstructure and mechanical properties of as-cast and age-hardened AlZnMg alloy.

2 Experimental methods

Al-10Zn-6Mg wt. % base alloys were produced using commercial pure Al (99.79 wt. %), Zn (99.99 wt. %), and Mg (99.90 wt. %) ingots with additions of Al-5Ti-B masteralloy rods at varying amounts to obtain final Ti content of 0.02, 0.05, and 0.25 wt. %. Starting materials were melted in a crucible furnace at 850 °C, degassed using argon gas for 1 min, and stirred immediately for 2 min. The melts were poured into a 170 x 170 x 15 mm^3 preheated metal mould followed with squeeze casting at 76 MPa for 10 min. Homogenisation was done at 400 °C for 4 h to uniform the microstructure of as-cast samples. Chemical composition of the Al-alloys determined by OES analysis using ARL 3460 OE Spectrometer is presented in Table 1. The alloys were then heat treated by T6 procedure consisting of solution treatment at 440 °C for 1 h followed with water quenching and ageing at 130 °C for 2-200 h. Samples for mechanical testing, microstructure, and phase characterizations were machined to prescribed test specifications.

Table 1. Chemical composition of as-homogenized Al-10Zn-6Mg-xTi wt. % alloys.

Sample	Zn	Mg	Ti	Fe	Al
A	10.318	6.741	0.004	0.172	*bal.*
B	9.701	6.734	0.018-0.0204	0.235	*bal.*
C	11.116	7.127	0.048-0.067	0.268	*bal.*
D	9.227	6.835	0.249-0.299	0.226	*bal.*

Mechanical testing was done through ASTM E 18 Rockwell B hardness testing and Charpy impact testing method. Zeiss-Primotech optical microscope and Zeiss SEM-EDS instrument were used for microstructural investigation on specimens prepared with standard metallographic preparation using 0.5 % HF etchant. Calculation of SDAS values were done on microstructure image of as-cast sample at 50 x magnification using ImagePro Analysis. Simultaneous Thermal Analysis (STA) was conducted to observe phase transformations, heat fluctuations, and alloys crystallinity as a function of time on temperatures on as-homogenized samples using STA 6000 Parkin Elmer at 30-750 °C with 10 °C/min scanning rate in accordance to ASTM E 168 standard.

3 Results and discussion

3.1 As-cast microstructures and hardness

Figure 1 presents the microstructures of Al-10Zn-6Mg alloys with various addition of Ti. Dendrites size decreased along with increasing content of Ti up to 0.25 wt. %. Microstructure of all samples, however, showed porosities occurring due to the existence of dissolved gas in the melting process. Eutectic phase β (Al_8Mg_5) was observed clearly inside the interdendritic phase. This is in agreement with the solidification sequence by Raghafan [12] which figured that Al-10Zn-6Mg alloy would form L + T at 480 °C and β at 448 °C, subsequently.

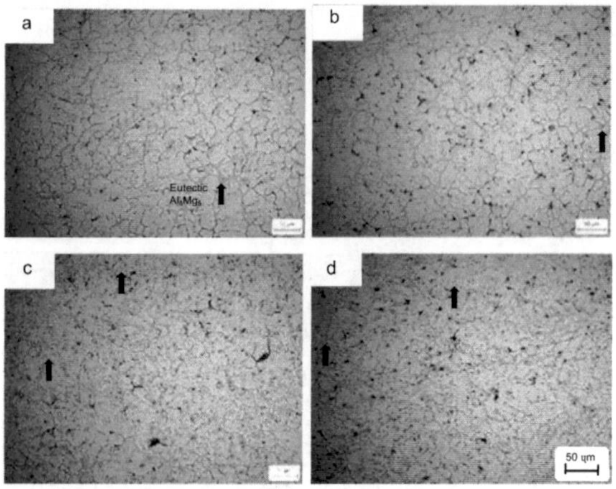

Fig. 1. Microstructure of as-cast Al-10Zn-6Mg alloy with : (a) 0, (b) 0.02, (c) 0.05, and (d) 0.25 wt. % Ti.

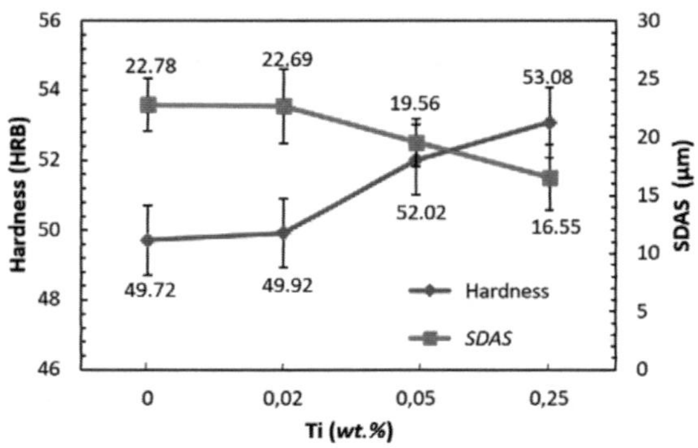

Fig. 2. The effect of Ti addition on hardness values and SDAS on Al-10Zn-6Mg-xTi alloys.

Ti addition affected the hardness and dendrites size of the alloy, as seen in Figure 2. As the addition of Ti increased, the SDAS values decreased while hardness of the alloy improved. Obtained SDAS values were 22.78, 22.69, 19.56, and 16.55 μm for the addition of 0, 0.02, 0.05, and 0.25 wt. % Ti, respectively. The hardness value increased gradually from 49.72 HRB to 49.92, 52.02, and 53.08 HRB, respectively for each Ti addition. Both results are related to each other. The addition of grain refiner promoted grain boundaries strengthening, thus improving the mechanical properties [3,4]. Grain refining action of Ti was achieved by the forming and distribution of $TiAl_3$ intermetallic phase in the melt, which became the nucleating place for α-Al. Another mechanism to note was the increasing of *Growth Restriction Factor* (GRF) due to the presence of solute Ti. The GRF value of the alloy escalated from 21 to 25.91, 33.28, and 82.4 at the addition of 0.02, 0.05, and 0.25 wt. % Ti respectively. Therefore, it can be inferred that grain refinement mechanisms of Ti in this alloy were α-Al nucleation and GRF escalation.

3.2 Effect of Ti on age hardening mechanism

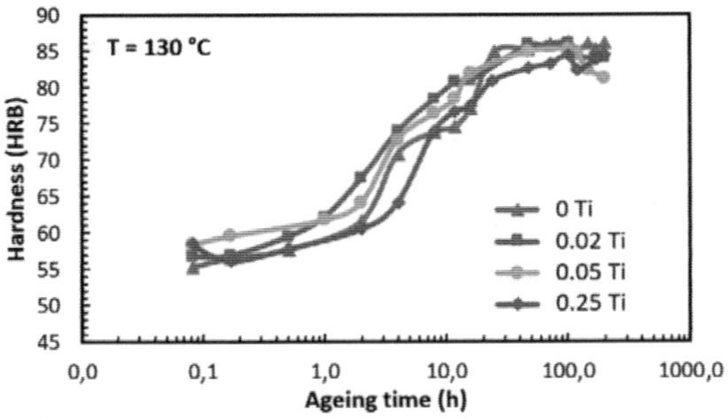

Fig. 3. Age hardening response of Al-10Zn-6Mg-xTi alloys at 130 °C for 200 h.

Age hardening response was observed in underaged, peakaged, and overaged conditions, of which hardness values can be seen in Figure 3. In the first 5 min, the alloys possessed underaged condition with hardness values of 55.3, 56.8, 58.3, and 58.4 HRB at addition of 0, 0.02, 0.05, and 0.25 wt. % Ti, respectively. Ti-doped alloys showed higher hardness values than those without. This is similar to a research by Gao, et. al. [6] which found that the presence of Ti accelerated initial ageing stage during artificial ageing at 120 °C, furthermore improved the hardness value. Peak ageing was observed at 102 h when hardness values of all samples significantly increased. However, sample without Ti addition showed higher hardness than Ti-added alloys. Alloys with 0, 0.02, 0.05, and 0.25 wt. % Ti had hardness values of 86.1, 85.9, 85.1, and 84.4 HRB, respectively. Moreover, the hardness values seemed to decrease as the Ti content was increasing. Therefore, it can be concluded that Ti did not induce either second phase or intermetallic phase formation at peakaged condition.

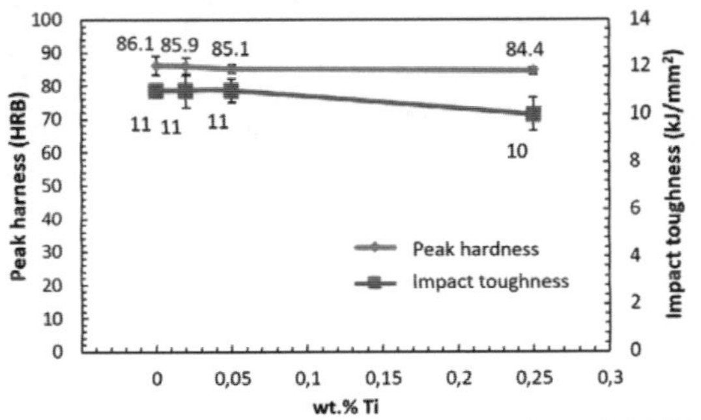

Fig. 4. Peak hardness and impact toughness values of Al-10Zn-6Mg-xTi at 130 °C.

Figure 4 illustrates the effect of 0, 0.02, 0.05, and 0.25 wt. % Ti addition on impact toughness and peak hardness of Al-10Zn-6Mg. It can be inferred that peak hardness and impact toughness were seemingly equal on all Ti variations; hence the addition of Ti contributed no significant effect on both aspects at peakaged condition. High level of hardness values at peakaged condition were a result of the formation of semi-coherent η' $MgZn_2$ precipitates in α-Al matrix. The precipitates distributed uniformly in the alloy during ageing furthermore inhibited dislocation movement due to deformation forces applied and improved the hardness value. Meanwhile, the addition of Ti, which contributed on reducing SDAS values, performed no effect on peak hardness. Instead, the formation of precipitates itself which took a big role on improving the mechanical properties. The images of fracture surface for each alloy with various Ti addition are shown in Figure 5. Fracture surface of alloys in as-aged condition were generally identified as brittle fracture. The images also show different grain fineness, where alloys with higher Ti content had finer grains.

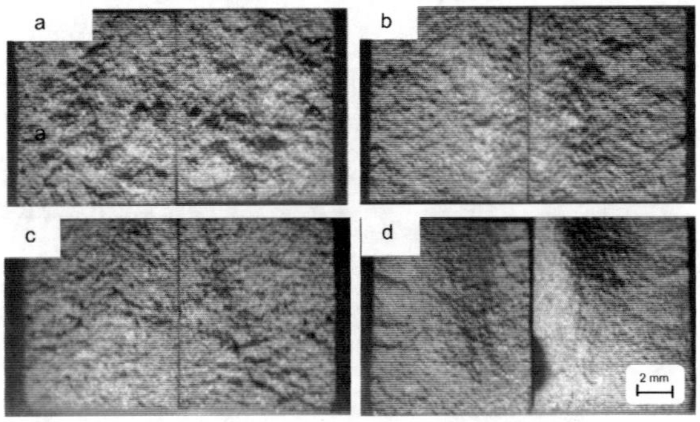

Fig. 5. Fracture surface of Al-10Zn-6Mg with addition of (a) 0, (b) 0.02, (c) 0.05, and (d) 0.25 wt. % Ti.

Microstructures of Al-10Zn-6Mg-xTi are shown in Figure 6. Interdendritic phase in white contrast and porosities in black contrast were observed in the structures. The number of interdendritic phase increased with more Ti addition. Porosities were formed due to the dissolution of hydrogen gas during casting process.

3.3 Phase transformation analysis with SEM/EDS

Figure 7 and Table 2 present the SEM images and EDX data of Al-10Zn-6Mg-xTi at peakaged condition. α-Al matrix, T ($Mg_{32}(Al,Zn)_{49}$), and β (Al_8Mg_5) were detected in Figure 7 (a-d). T phase was observed in cubic shape due to a low ratio of Zn:Mg. Interestingly, point 12 shows spider web-shaped cubic β phase inside the interdendritic structure. T and β phases were formed during the initial process of solidification. The low ratio of Zn:Mg caused η ($MgZn_2$) phase to be undetected in all alloy compositions. $TiAl_3$ was supposed to form in nanoparticle size that the morphology could not be observed. C and O_2 content was found in the alloy due to residual Al-SiC polishing cloth and paste (Al_2O_3).

3.4 Phase transformation analysis with STA

Phase transformation of the alloys can be seen through the endothermic and exothermic peak in Figure 8 (a-d). Reaction occurring in Al-10Zn-6Mg-xTi was SSSS → GP zone → η' → η [13]. Table 3 shows the formation (exothermic) and diffusion (endothermic) temperatures of the alloys due to the heating process. Khrisna, et. al. [13] observed that the GP zone, η', and η developed at 20-120 °C, 120-150 °C, and 150-300 °C, respectively, while diffusion of each phases took place at 50-150 °C, 200-250 °C, and 300-350 °C, respectively, which are appropriate with findings in this work.

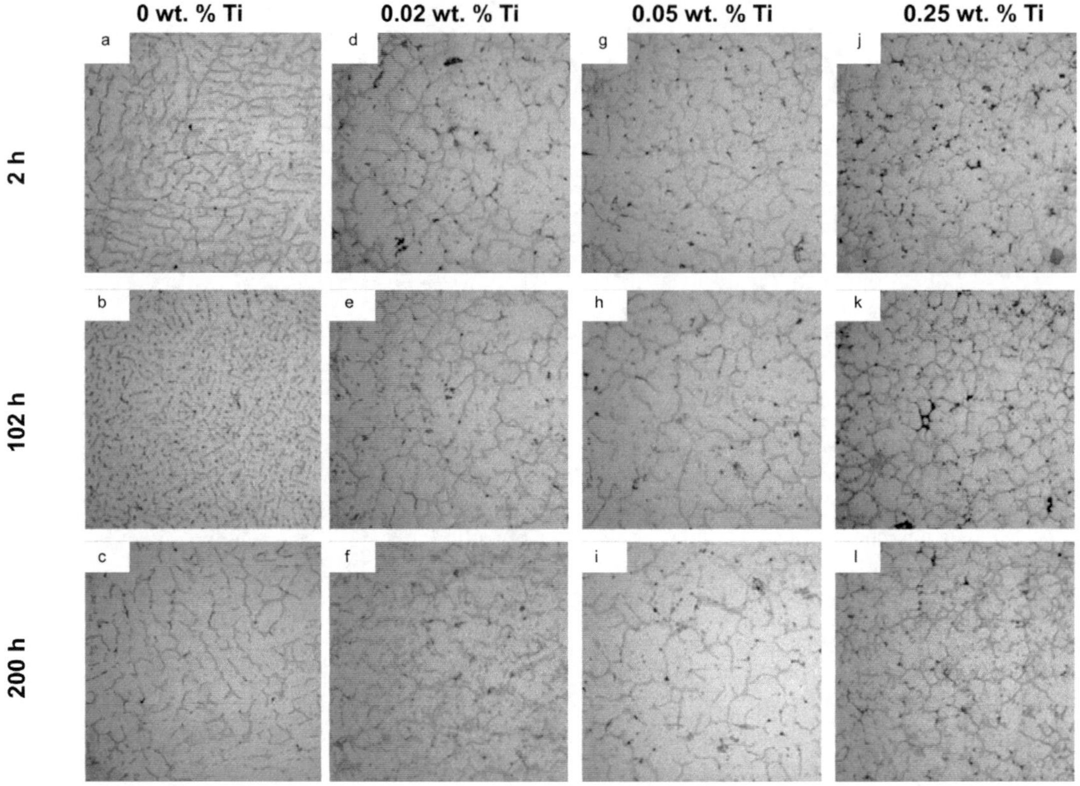

Fig. 6. Microstructure of Al-10Zn-6Mg alloy with: (a-c) 0, (d-f) 0.02, (g-i) 0.05, and (j-l) 0.25 wt. % Ti at underaged (2 h), peakaged (102 h), and overaged (200 h) condition with ageing temperature of 130 °C.

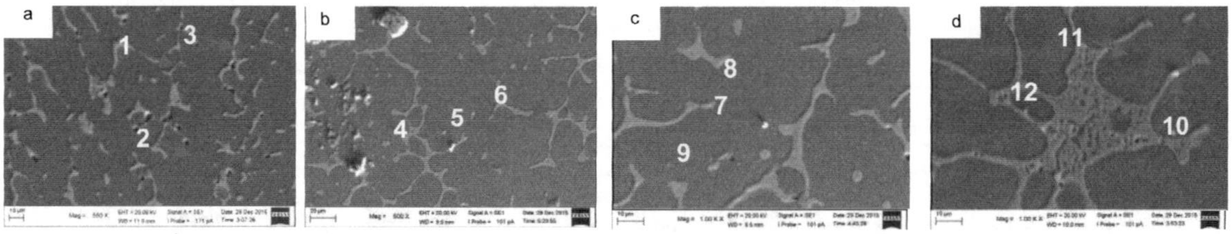

Fig. 7. SEM images of Al-10Zn-6Mg with addition of (a) 0, (b) 0.02, (c) 0.05, and (d) 0.25 wt. % Ti in peakaged condition.

Table 2. EDX characterization on Al-10Zn-6Mg-xTi

Mark	Element content (at. %)						Possible phase
	Al	Zn	Mg	Ti	C	O	
1	Bal	12.65	15.9	-	18.8	3.63	$Mg_{12}(Al,Zn)_{49}$
2	Bal	1.72	5.13	-	15.9	2.67	α-Al
3	Bal	1.84	5.22	-	14.9	2.87	Al_8Mg_5
4	Bal	0.68	2.23	1.83	45.8	36.1	$Mg_{12}(Al,Zn)_{49}$
5	Bal	-	1.52	0.05	46.3	37.9	α-Al
6	Bal	-	1.15	-	49.4	36.3	Al_8Mg_5
7	Bal	0.10	4.41	-	39.4	20.6	$Mg_{12}(Al,Zn)_{49}$
8	Bal	-	2.69	-	47.6	24.4	α-Al
9	Bal	-	4.17	-	29.9	9.89	Al_8Mg_5
10	Bal	13.09	20.9	-	24.1	6.80	$Mg_{12}(Al,Zn)_{49}$
11	Bal	-	4.24	-	29.5	13.3	α-Al
12	Bal	-	8.96	0.06	31.9	47.7	Al_8Mg_5

Table 3. Phase formation and diffusion temperature of Al-10Zn-6Mg-xTi.

wt. % Ti	Temperature °C		
	GP zone	η'	η
Formation			
0	40	120	200
0.02	40	150	190
0.05	40	120	210
0.25	40	180	360
Diffusion			
0	70	180	230
0.02	100	160-180	220
0.05	100	140	240
0.25	140	300	380

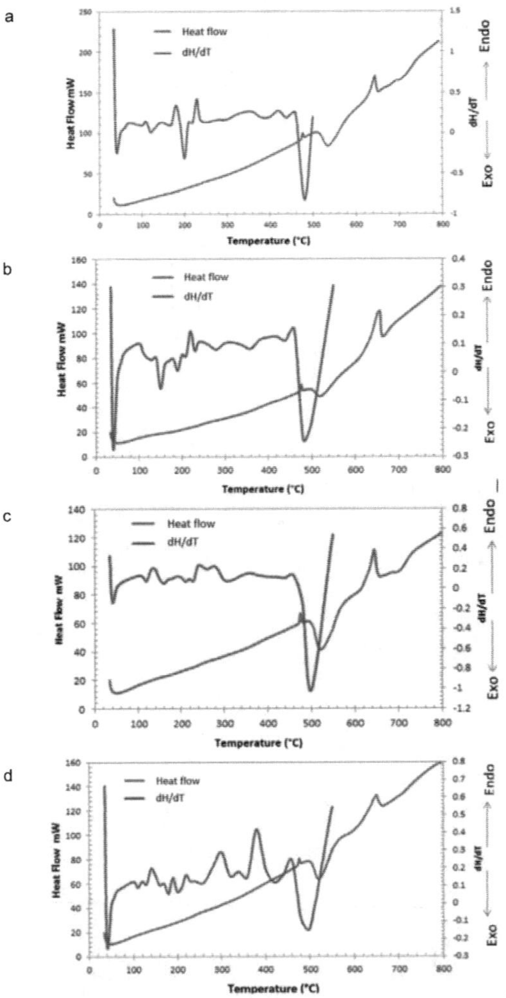

Fig. 8. Heat flow curves and their derivatives of Al-10Zn-6Mg with addition of (a) 0, (b) 0.02, (c) 0.05, and (d) 0.25 wt. % Ti.

Results of STA analysis in this work showed that formation of GP zone, η', and η occurred at 40 °C, 120-180 °C, and 190-360 °C, respectively. The diffusion of GP zone took place at 70-140 °C, η' at 140-300 °C, and η at 220-380 °C. In addition, the heat flow curves showed two endothermic peaks. The first one was indicated as β phase diffusion, which occurred at 476.67, 475.43, 475.25, and 475.85 °C for alloy with 0, 0.02, 0.05, and 0.25 wt. % Ti, respectively. β phase was formed at 447 °C. Additionally, the second endothermic peak presented the α-Al phase diffusion for each alloy with 0, 0.02, 0.05, and 0.25 wt. % Ti which were at 644.06, 656.53, 645.05, and 649.23 °C. It was lower than 660 °C due to alloying content which decreased the melting point of base metal. In summary, Ti addition gave no effect on the temperature of phase transformation.

4 Conclusions

This work investigated the microstructure, phase transformations, and mechanical properties of aged Cu-10Zn-6Mg alloys with addition of 0.02, 0.05, and 0.25 wt. % Ti. Results are summarized as follows:

1. Addition of 0.02, 0.05, and 0.25 wt. % Ti increased the hardness of as-cast Al-10Zn-6Mg wt. % alloys from 49.72 to 49.92, 52.02, and 53.08 HRB, simultaneous to the reduction of its SDAS value from 22.78 to 22.68, 19.56, and 16.55 μm and a rise in GRF from 21 to 25.91, 33.28, and 82.4, respectively.
2. Ti did not contribute to the peak hardness of Al-10Zn-6Mg wt. % alloys which were found to be 86.1 HRB and 85.9, 85.1, and 84.4 HRB at 0.02, 0.05, and 0.25 wt. % Ti additions, respectively, proportional to their impact numbers, ranging between 10-11 kJ/m^2.
3. Microstructural investigation on Al-10Zn-6Mg-xTi wt. % alloys showed the presence of primary α-Al matrix, spider web-shaped β (Al_8Mg_5), and bright interdendritic T ($Mg_{32}(Al,Zn)_{49}$) phases.
4. Phase transformation during age hardening of Al-10Zn-6Mg-xTi wt. % alloys was identified as an exothermic formation of η' following the order of GP Zone → η → η' ($MgZn_2$) and its endothermic diffusion. The formation of β (Al_8Mg_5) was observed at 447 °C while its diffusion occurred at approx. 475 °C.

Acknowledgements

This work was supported by Directorate of Research and Community Services, Universitas Indonesia through Hibah PITTA (*Publikasi Internasional Terindeks untuk Tugas Akhir Mahasiswa*) UI 2018.

References

1. M. Syahid, B.T. Sofyan, J. Adv. Mat. Res. **789**, 324-329 (2013)
2. F.H. Kurniawan, *Pengaruh Penambahan 0.067, 0.081, dan 0.115 wt. % Ti thd Karakteristik Paduan AC4B Hasil LPDC* (Undergrad. Thesis, Universitas Indonesia, 2008)
3. B.T. Sofyan, D.J. Kharistal, L. Trijati, K. Purba, R. Susanto, Mat. Des. **31**, S-36-S43 (2010)
4. P.I. Lestari, *Studi Pengaruh Penambahan TiB terhadap Karakteristik Komposit Al-10Zn-6Mg-3Si Berpenguat ZrO_2 Hasil Squeeze Casting* (Undergrad. Thesis, Universitas Indonesia, 2009)
5. G.K. Sigworthh, T.A. Kuhn, *Grain Refinement of Aluminium Casting Alloys* (AFS, 2007)
6. T. Gao, Y. Zhang, X. Liu, Mat. Sci.&Eng. A **598**, 293-298 (2014)
7. B.T. Sofyan, L. Basuki, Proc. 11th Int. Conf. Quality in Research, 832-839 (2009)
8. B.T. Sofyan, *Pengantar Material Teknik* (Penerbit Salemba, 2009)
9. C.A. Grove, G. Judd, Met. Trans. **4**, 1025 (1973)

10. M. Murayama, K. Hono, Acta Mat. **47**, 1537-1548 (1999)
11. S.K. Maloney, K. Hono, I.J. Polmear, S.P. Ringer, Scripta Mat. **41**, 1031-1038 (1999)
12. V. Raghafan, J-Alloy Phase Diagrams. **28**, 203-208 (2007)
13. K.G. Khrisna, K. Sivaprasad, K. Venkateswarlu, K.C.H. Kumar, J. Mat. Sci. Eng. **535**, 129-135 (2012)

MATEC Web of Conferences **186**, 02010 (2018) https://doi.org/10.1051/matecconf/201818602010
ICEMP 2018

Effect of Mechanical Load on Defects Level in Soft Ferrite Ceramics

A.B. Petrova[1] , A.V. Malyshev[1] and A.P. Surzhikov[1]

[1]Tomsk Polytechnic University, School of High-Energy Physics, 634050 30, Lenin Avenue, Tomsk, Russia

Abstract. The article presents the results of a study of the effect of mechanical load on the temperature dependence of the initial permeability of LiTiZn soft ferrite ceramics. Regimes of mechanical load were created by a steel non-magnetic ring with screws. In this work, four levels of mechanical load were investigated: without load, with 1, 2 and 3 screws. For obtaining temperature dependences and exclude the influence of prehistory on the initial permeability, the samples were heated to a temperature exceeding the Curie point by 50 degrees, after which they slowly cooled to room temperature. The defects level was determined by the ratio of the parameters β / α of the phenomenological expression, which describes the experimental temperature dependences. It is shown that as the mechanical load increases, the defects level of ferrite ceramics increases, and after load removing, it returns to its original level.

1 Introduction

In modern radio electronics, soft ferrites find wide application for manufacturing magnetic amplifiers, inductors, transformer cores, rotors and stators of high-frequency motors [1, 2]. The most widely used is ferrites synthesized by ceramic technology from a mechanical mixture of oxides or carbonates. The process of ferrites preparation on ceramic technology includes several stages, the purpose of which is the homogenization of the oxide mixture. The advantage of ceramic technology is the low cost and relative simplicity of the technological cycle. However, homogenization techniques used in ceramic technology lead to the appearance of inhomogeneities of a different type, which can have a negative effect on the magnetic characteristics of ferrite [3]. Different methods such as method of measure the true heat capacity, measure the temperature variation of magnetization in the temperature range including the Curie point, and micro-X-ray spectral analysis is used for registration defects and evaluation the products homogeneity from soft ferrite ceramics [4]. Studies of the temperature dependences of the magnetic parameters of ferrites are presented in the works [5, 6], an analysis of the influence of the ferrite homogeneity on the sharpness of the drop in the temperature curves is made. The disadvantages of these methods are low accuracy, laboriousness, the inability to quantify the changes in chemical composition. For solving these problems, in the Tomsk Polytechnic University is developing a method based on measurements the temperature dependence of the structurally sensitive characteristic of soft ferrites - the initial permeability.

The purpose of this work is approbation the method for evaluating the defects level of ferrite ceramics, based on measuring the temperature dependences of the initial permeability of samples with the mechanical load.

It was expected that an increase of the mechanical load and associated elastic stresses would lead to an increase in the defects level (β / α) of ferrite ceramics.

2 Experimental techniques

The object of the study is ferrite ceramics, manufactured using classical ceramic technology. Chemical formula: $Li_{0.649}Fe_{1.598}Ti_{0.5}Zn_{0.2}Mn_{0.051}Bi_{0.002}O_4$. The test samples were manufactured by the Ferrit-Domain Co. under production conditions at T=960° C, t=8 h. After sintering, the samples had the toroidal form with an external diameter D=20.5 mm, an internal diameter d=10.9 mm, a height h=3.5 mm.

Samples density and porosity were determined by hydrostatic weighing on an analytical balance Shimadzu AUW 220D.

The grain morphology was studied with a scanning electron microscope Hitachi TM-3000. The average grain size was calculated using the intercept method.

The saturation magnetization (Ms) was measured at room temperature on a magnetometer with a vibrating sample with a maximum field of 0,8 MA/m.

An installation based on an automatic bridge LCR-819 and a measuring cell with a built-in heater was built for the measurements. Measuring cell is shown in Figure 1. The measuring cell consists of a stainless steel cylinder (1), which is closed by a fluoroplastic lid (2). A device holding sample (3) and thermocouple (4) is fixed on the lid. Also on the lid, there are holes for the output of conductors (5) from the winding of sample (7).

Mechanical load is created by a specially designed ring (8) of solid non-magnetic steel with three holes for screws DIN 553. In the lower part of the cylinder there is a pressed copper cup (12), at the bottom of the cup there is a heater (9). The upper part of the cup is covered with an aluminum heat shield (6) to increase the uniformity of heating. All elements are located in an aluminum case (11). The body has connectors that allow connecting the ground wire and the wires that supply the heating element. The body of the measuring cell is thermally insulated with powder of light-weight fireclay bricks 10.

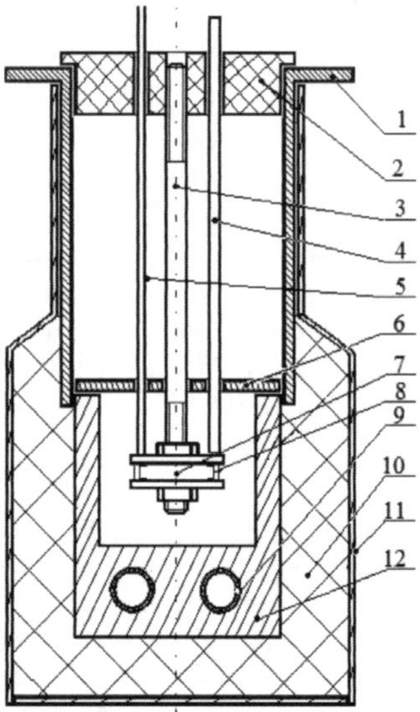

Figure 1. Measuring cell: cylinder (1), fluoroplastic lid (2), device holding sample (3), thermocouple (4), wire outlet (5), heat shield (6), sample (7), steel ring (8), heater (9), fireclay powder (10), aluminum case (11), copper cup (12).

Measurement of the inductance L was carried out on a low-frequency meter LCR-819 in an electromagnetic field with a frequency of 1 kHz with strength of 8 A/m. Installation is shown in Figure 2.

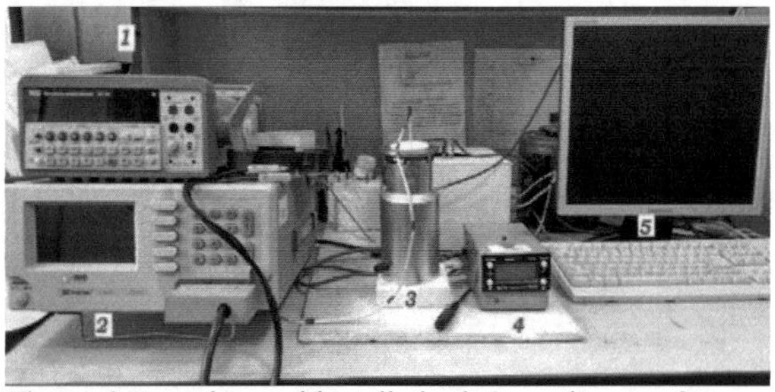

Figure 2.Experimental installationl: 1 – voltmeter, 2 – digital meter L, C, R, 3 – measuring cell, 4 – heater, 5 – computer.

Ferrite toroid had a single-layer winding. Winding was made of thin copper wire with fluoroplastic insulation.

A non-magnetic steel ring with non-magnetic steel screws was used for creating a mechanical load on sample. Samples were placed into this ring as shown in Figure 3. Mechanical load was created by alternately screwing screws into the sample. Five measurement regimes were chosen: off-load, with one screw, with two screws, with three screws, and after removing the mechanical load.

Figure 3. Soft ferrite sample with winding, placed in the steel ring with screws.

The toroidal sample was placed in the measuring cell shown in Figure 1. The sample was measured with a slow cooling of the sample from a temperature 50 degrees above the Curie point using an installation based on an automatic bridge LCR-819 and a voltmeter B7-78 / 1. Temperature control was carried out using a thermocouple of the chromel-alumel group (type K). The measuring junction of the thermocouple was pressed directly to the sample. The inductance data L from the LCR-819 and the temperature T were supplied as a text file to the computer using the LabVIEW software package. Calculation of initial permeability values and their mathematical processing took place in the program Origin 9.

The values of the initial permeability were calculated from the formula:

$$\mu_i = \frac{L \cdot 10^7}{2 \cdot h \cdot N^2 \cdot \ln(\frac{D}{d})}; \tag{1}$$

where L – coil inductance (H); N – number of coil turns; h – samples height (m); D, d – external and internal diameters of samples (m).

Phenomenological expressions (2, 3) were used for the mathematical analysis of the experimental temperature dependences of the initial permeability [7]:

$$\mu_i^* = \frac{1+\mathrm{x}}{1+N\cdot\mathrm{x}} \tag{2}$$

$$x = \left[\frac{\left(1-\frac{T}{T_c}\right)^{\delta}}{\alpha(1-\frac{T}{T_c})^{\gamma}+\beta}\right] \tag{3}$$

In expression (2) the coefficients α, β, and γ are determined by the relations:

$$\alpha = \frac{K_1(0)}{M_s^r(0)} \qquad (4)$$

$$\beta = \frac{\lambda_s(0)}{M_s^r(0)} \cdot \sigma \qquad (5)$$

$$\delta=(r-n)f, \qquad (6)$$

$$\gamma=(m-n)f. \qquad (7)$$

Where σ – the average level of elastic stresses, N_f – demagnetization factor, and *g* and *r* are specified according to the selected magnetization reversal mechanism [7]. In our work indicators *g*, *r=2* was chosen according to the Smith and Wijn model, which is used for polycrystalline ferrites [4].

With positive values of γ and δ functions (2, 3) describe an asymmetric peak shape, characteristic for typical dependences $\mu_i(T)$.

The least-squares method (the Levenberg-Marquardt algorithm) of the Origin 9 software was used to process the experimental temperature dependences of initial permeability.

3 Experimental results

Figure 4 shows the X-ray diffraction patterns of a ferrite ceramics sample obtained under the sintering regime T=960 °C, t=8 h. Also in this figure there are the values corresponding to the diffraction peaks. The X-ray diffraction patterns of the test sample confirmed the formation of a single-phase homogeneous cubic spinel structure.

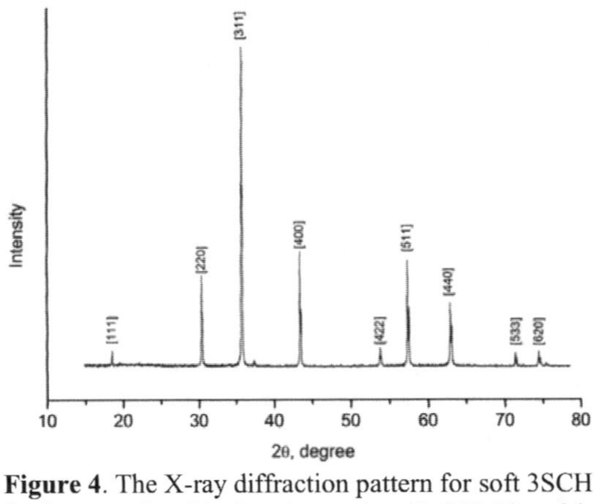

Figure 4. The X-ray diffraction pattern for soft 3SCH18 ferrite ceramic with sintering regime T=960 °C, t=8 h.

A micrograph obtained by scanning electron microscopy on Hitachi TM-3000 is shown in Figure 5.

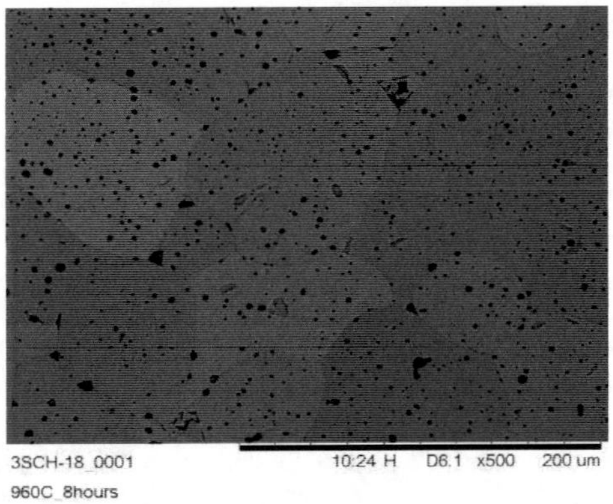

Figure 5. SEM micrograph of soft 3SCH18 ferrite ceramic with sintering regime T=960 °C, t=8 h.

Figure 5 presents that the investigated ceramics is characterized by a polycrystalline structure with well-formed grains.

Table 1 presents the average grain size, density, and porosity of soft ferrite ceramics 3CH18.

Table 1. Parameters of soft 3SCH18 ferrite ceramic with sintering regime T=960 °C, t=8 h.

Average grain size (μm)	Density (g/cm^3)	Porosity (%)	M_s (G)
47,2	4,3	4,4	1930

As can be seen from Table 1, in spite of the large grain size, the test ferrite has a high density, which reaches 96% of the X-ray density for this material and a sufficiently low porosity.

The temperature dependences of the initial permeability soft ferrite ceramics samples with different mechanical load regimes are shown in Figure 6.

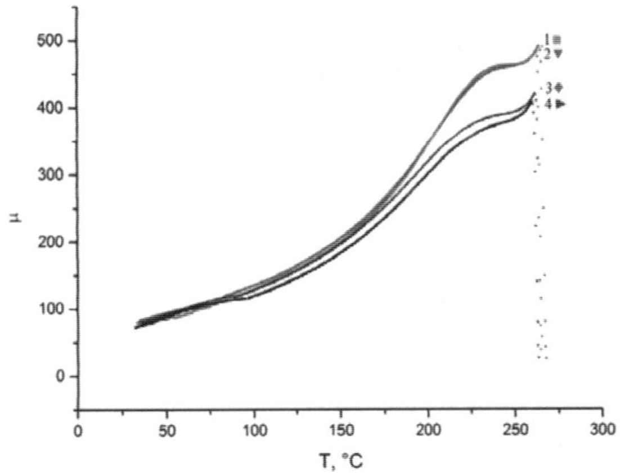

Figure 6. The temperature dependences of the initial permeability soft ferrite ceramics samples with different mechanical load regimes: 1 – off-load; 2 – with 1 screw; 3 – with 2 screws; 4 – with 3 screws.

As can be seen from Figure 6 the initial permeability near the Curie point for all ferrite samples has the sharp decrease. As shown in the works [5, 6, 8], this behavior of the curve is characteristic of the single-phase, the chemically homogeneous structure of ferrite ceramics. Thus, the test soft ferrite ceramic has good homogeneity.

The obtained temperature dependences are characterized by two stages of growth. The first stage is characterized by a gradual increase in the initial permeability from room temperature to 250 °C. The second stage is characterized by the Hopkinson peak, which is a sharp increase with the pass of an acute maximum and a sharp fall in the values of the initial permeability near the Curie point. This behavior is associated with the domains transition to a superparamagnetic state, which is inherent in the single-domain state in the case when the domain size and the average grain size coincide, for example, in the case of a nanoparticle [9-11].

The highest maximum of the dependence of the initial permeability on temperature is characteristic for the sample off-load mechanical load. The value of the maximum of the initial permeability decreases with increasing load on the test sample. Therefore, this parameter can be used to estimate the level of mechanical stress and the associated level of elastic stresses in the material.

Table 2 presents the results of mathematical processing of the temperature dependences of the initial permeability of soft ferrite ceramics samples with different levels of mechanical load using the phenomenological expression (2).

Table 2. Results of mathematical processing

Mech. load regime	α	β	N_f	T_c	β/α
Off-load	0,132	6.28E-6	0.0020	266.32	4.8E-5
1 screw	0,120	5.47E-5	0.0024	264.76	4.6E-4
2 screws	0,121	1.54E-5	0.0020	265.67	1.3E-4
3 screws	0,127	1.09E-4	0.0026	268.78	8.6E-4
After a load	0,123	8.19E-6	0.0020	273.46	6.7E-5

As can be seen from Table 2, with increasing mechanical load, the parameter β / α increases due to the increase in elastic stresses and defects level in the sample. After removing the mechanical load, the β / α ratio returned to its original value. The demagnetizing factor N_f is similar for all regimes of mechanical load and practically does not change.

4 Conclusions

Thus, in this paper, the influence of the mechanical load on the temperature dependences of the initial permeability of ferrite ceramics was investigated.

It is shown that as the mechanical load increases, the structurally sensitive parameter of the phenomenological expression β / α increases, which is caused by the growth of elastic stresses and defects level in the sample. After the load is removed, the parameter β / α returns to the original level, which indicates that there are no irreversible changes caused by the mechanical load. The maximum of the temperature dependence of initial permeability can be used for estimates of the mechanical load level and the associated elastic stress level in the material.

Thus, in this paper, a method for estimating the defectiveness of soft ferrite ceramics samples subjected to mechanical load was approbated.

Acknowledgments

The research is funded by The Ministry of Education and Science of the Russian Federation as part of the ''Science'' Program, Project 3.4937.2017.

The experimental calculations were carried out at Tomsk Polytechnic University within the framework of Tomsk Polytechnic University Competitiveness Enhancement Program Grant.

References

1. R. Valenzuela, Phys. Res. Inter., **2012**, 1-9, (2012).
2. L. Jaswal, B. Singh, Journal of Integrated Science and Technology, **2**, 69-71, (2014).
3. B.K. Ostafiychuk, L.S. Kaykan, J.S. Kaykan, B.Y. Deputat, O.V. Shevchuk, Nano. Res. Let., **12**, 237, (2017).
4. J Smith, H.P.J. Wijn, *Philips Technical Library,* Eindhoven, Netherlands, 369, (1959).
5. R. Parvin, A.A. Momin, A.K.M. Akther Hossain, J. Mag. and Magn. Mat., **401**, 760-769, (2016).
6. J.S. Ghodake, T.J. Shinde, R.P. Patil, S.B. Patil, S.S. Suryavanshi, J. Mag. and Magn. Mat., **378**, 436-439, (2015).
7. A.B. Petrova [et al] IOP Conf. Ser.: Mater. Sci. Eng., **289**, 012042, (2018).
8. S. Islam, K.H. Maria, S. Choudhury, M. Hakim, D.K. Saha, Turk. J. Phys, **36**, 253-259, (2012).
9. M. Marimuthu, S.K. Krishnaswamy, N. Aparnadevi, C. Venkateswaran, Phys. Status Solidi A., **212**, (2015).
10. H. Sözeri, A. Baykal, and B. Ünal, Phys. Status Solidi A., **209**, 2002-2013, (2012).
11. Q. Fang, H. Cheng, K. Huang, J. Wang, R. Li, and Y. Jiao, J. Magn. Magn. Mater. **294**, 281-286, (2005).

AUTHOR INDEX

EDP Sciences – Web of Conferences
17, Avenue du Hoggar
Parc d'Activité de Courtabouf
BP 112
F-91944 Les Ulis Cedex A
France

ISBN 978-1-5108-6860-1

CURRAN ASSOCIATES INC.
proceedings.com

21st Conference on Power System Engineering 2022

MATEC Web of Conferences Volume 367 (2022)

Pilsen, Czech Republic
7-9 September 2022

edp sciences